BACTERIOLOGY RESEARCH DEVELOPMENTS

RECENT DEVELOPMENTS IN *ENTEROBACTER* RESEARCH

Bacteriology Research Developments

Additional books and e-books in this series can be found on Nova's website under the Series tab.

BACTERIOLOGY RESEARCH DEVELOPMENTS

RECENT DEVELOPMENTS IN *ENTEROBACTER* RESEARCH

GRÉGOIRE ROUX
EDITOR

NOTICE TO THE READER

Additional color graphics may be available in the e-book version of this book.

Library of Congress Cataloging-in-Publication Data

Names: Roux, Grégoire, editor.
Title: Recent developments in enterobacter research / Grégoire Roux, editor.
Description: New York : Nova Science Publishers, [2020] | Series: Bacteriology research developments | Includes bibliographical references and index. |
Identifiers: LCCN 2020041243 (print) | LCCN 2020041244 (ebook) | ISBN 9781536186154 (paperback) | ISBN 9781536186888 (adobe pdf)
Subjects: LCSH: Enterobacter sakazakii. | Enterobacter.
Classification: LCC QR82.E6 R43 2020 (print) | LCC QR82.E6 (ebook) | DDC 579.3/4--dc23
LC record available at https://lccn.loc.gov/2020041243
LC ebook record available at https://lccn.loc.gov/2020041244

Published by Nova Science Publishers, Inc. † New York

CONTENTS

PREFACE

Enterobacter is a relatively lesser known member of family Enterobacteriaceae with at least fifteen species more or less involved in different human infections. As little knowledge is available about its pathogenicity and virulence factors, this compilation discusses the factors and genes involved. The plants and microbes with antimicrobial potential in the synthesis of nanoparticles for the control of pathogenic Enterobacteriaceae are highlighted, and the authors discuss how, to overcome the complexities regarding antibiotic therapies and multiple drug resistance, there is a dire need to develop some novel clinical approaches and strategies. The factors causing multi-drug resistance are highlighted, including under or overuse of antibiotics, prolonged use of antibiotics, poor infection control, poor hygiene and sanitation.

Chapter 1 - *Enterobacter* (genus) relatively lesser known member of family enterobacteriaceae, yet with at least fifteen species more or less involved in different human infections. A very little knowledge is available about the pathogenicity and virulence factors of different species of *Enterobacter*. These Gram-negative bacteria possess endotoxins and have complex pathogenetic mechanism as sustain quite a number of putative virulence factors. But still their role in infections in host is not well studied. Some studies showed that *in vitro, Enterobacter* spp. produce some virulence factors after adhesion to host epithelial cells. These factors include enterotoxins, α-hemolysin and thiol-activated cytotoxins (pore-forming). In several Gram-negative bacteria including *Enterobacter* spp., the type III secretion system (TTSS) has proved to be very important with various proteins. This system delivers virulence factors (toxins) in host cells. Thus TTSS plays an important role in pathogen-host interactions. The presence of the TTSS can be used as a general indicator of bacterial virulence. *E. cloacae* a well known pathogenic species mostly isolated from clinical samples, has 27% TTSS genes, a clear indication that this bacteria can damage epithelial cells and phagocytes within the host. *E. cloacae* pathogenesis is majorly contributed by cytotoxic enterotoxins produced by 2-mercaptoethanol and the TTSS. *E. cloacae* strains may induce apoptosis of HEp-2 cells in host as primary strategy to spread systemic infection. Another important factor is the biofilm formation by producing proteinaceous extracellular fibers called curli fimbriae. These fimbriae help bacteria to bind with a variety of host proteins and also adhesion and invasion. These curli fimbriae are encoded by curli gene in *E. cloacae* and other species of *Enterobacter*. In this chapter the virulence factors involved in pathogenesis and the genes associated with these factors in *Entrobacter* spp. shall be discussed.

Chapter 2 - *Introduction:* Members of family Enterobacteriaceae are pathogenic and offer potential health risks to human and animals. These members form pathogenesis by using different biochemical and molecular strategies to cope with host defense system. The use of classic approaches like antibiotics besides being expensive to implement poses serious threats to ecosystems. The efficacy of reported use of different nanoparticles as antimicrobial agent is one of the novels and yet to explore strategies for control of pathogenic microorganisms from family Enterobacteriaceae. Conventionally physical and chemical methods are used for the synthesis of diverse types of nanoparticles. Although these methods are giving promising results but there are some limitations regarding the use of methods as they are not ecofriendly and use toxic chemicals. Therefore, interest in the synthesis of nanoparticles by using biological sources has been amplified. Biological methods used for the synthesis of nanoparticles are not only cost effective, can be easily scaled up but also environmentally friendly. Biological sources like plant and microbes including bacteria, algae, and fungi are well known for their potential to produce diverse range of nanoparticles.

Conclusion: Current review will highlight the plants and microbes having antimicrobial potential for using in the synthesis of nanoparticles for the control of pathogenic Enterobacteriaceae in safe way.

Chapter 3 - During the last few decades, the antimicrobial resistance due to overuse and misuse of antibiotics has become significant and springing as a grave risk to global health concerns. The rising rate of multiple drug resistance (MDR) is reducing the reliability of first-line traditional available pharmaceutics like cephalosporins, penicillins and fluoroquinolones for the treatment of community and hospital-acquired (nosocomial) infections caused by the pathogens of family Enterobacteriaceae. High rates of morbidity and mortality along with the prolonged hospitalization are common fates in such bacterial contagiousness. To overcome the complexities regarding the antibiotic therapies, MDR and to combat the pathogens, there is a dire need to develop some novel clinical strategies.

In this perspective, the world scientific community suggests the combined use of multiple drugs that target peptidoglycan synthesis as compare to monotherapy, which shows promising results in combating the infections caused by Enterobacteriaceae. Moreover, various antibiotics like polymyxins, fosfomycin and tigecycline (semisynthetic glycylcycline) have also provided the efficacious therapeutic alternatives to clinicians for critically ill hosts. Polymyxin B and colistin are remerged as a last resort therapeutics for the treatment of nosocomial infections caused by superbugs (pathogens which are almost resistant to every existing antibiotic). Furthermore, cationic antimicrobial peptides via autolysis and enzyme synthesis blockage, peptidomimetics and foldamers (non-toxic small peptides with amphiphilic nature) via membrane lysis, siderophores and sideromycins via depletion of iron needed for the growth of bacteria, efflux pump inhibitors via high intercellular drug persistency and therapeutic antibodies via phagocytosis to neutralize the different strains like *Enterococcus, Escherichia coli* and *Klabsiella pneumonia* (Enterobacteriaceae).

In addition to the on-deck antibiotics, clinicians and medicinal experts are intended to revive the use of old and abandoned drugs like temocillin and fosfomycin with modernistic and different combinations and strategies to treat the infections of Enterobacteriaceae. Moreover, bacteriophages (natural predators of bacteria) are again in use for the treatment of different MDR pathogenic organisms like *Pseudeomonas aeruginosa* and *Staphylococcus aureus*. As the MDR arose due to the presence of lactamases in the pathogens, the global health sector is in pipeline to design the β-lactam/β-lactamase inhibitors with significant and promising *in vitro* activity to combat the MDR pathogens. Further deep analysis with the clinical approach is needed alongwith the prolific and practical strategies to formulate the suitable drug candidates to cope up with the battle against the emerging MDR.

Chapter 4 - The *Enterobacter cloacae* Complex (ECC) as a human pathogen has wide geological spread causing infections of urinary tract, respiratory tract, intra-abdominal and bacteremia. Numerous lineages show multiple resistance traits like Ampicillin C (AmpC) and clonal lineages of ECC with high epidemic potential can be linked with antibiotic resistance evolution. Some examples of antibiotic resistant traits in ECC are Extended-spectrum β-lactamases (ESBLs), Metallo-β-lactamases (MBLs), Cephalosporinases and Carbapenemases. In hospital environment, surfaces transmit several key healthcare-associated pathogens including ECC. Other factors causing multi-drug resistance include under or overuse of antibiotics, prolonged use of antibiotics, poor infection control, poor hygiene and sanitation. Multi-drug resistance of ECC has become a major challenge for physicians as it complicates treatment of infections; many people appear non-compliant with full regime of treatment due to development of drug resistance in bacteria. Molecular studies reveal that drug resistance has intrinsic (chromosomal) and extrinsic (plasmid) mutation mechanisms. Mutated resistance carrying bacterial strains survive and replicate evolving into the dominant types. Resistance against broad spectrum cephalosporins and aztreonam is apparently caused by hyper production of constitutive AmpC due to multiple regulatory mutations. Multiple resistant genes for production of ESBLs and MBLs and quinolone resistance-determining region (QRDR) gene mutations have been identified to cause resistance against fluoroquinolones. There is a major threat that multi-drug resistance in ECC can lead the world to a post-antibiotic era where infection or injury can result into non-treatable or fatal complications and it can have a drastic impact on the global economy through increase in disease burden as well.

Chapter 5 - In recent times, the taxonomy of genus *Enterobacter* has gone through significant revision. The recent classification is based on DNA-DNA hybridization technique and 16S rRNA sequence analysis. Now this genus contains fifteen named species. Many members of this genus have already been transferred to other genera of Enterobacteriaceae family. These genera are; *Klebsiella*, *Serratia*, *Hafnia*, and *Pantoea*. This transfer has significant current genetic evidence in support. Transfer of *Enterobacter aerogenes* to genus *Klebsiella* and reclassification of a number of geno-species in the E. *cloacae* complex of organisms to different named species are good recent examples. On the other hand, some members of genus *Erwinia* have been transferred to genus *Enterobacter*. The *Enterobacter* species most commonly isolated from human sources are; *E. cloacae, E. aerogenes*, and to a lesser extent, *Pantoea* (previously *Enterobacter*) *agglomerans* group, *Cronobacter* spp., *E. gergoviae,* and *E. asburiae*. *E. cloacae* is most common clinical isolate from human samples. Many newly named species (formerly called *E. cloacae* group) are difficult to differentiate only by morphological or biochemical test methods. So, advance molecular techniques are used for their classification in novel way. Matrix-assisted laser desorption/ ionization–time of flight (MALDI-TOF) mass spectrometry is very helpful in identification of gram-negative bacilli and differentiation among *Enterobacter, Cronobacter,* and *Pantoea* species.

In: Recent Developments in *Enterobacter* … ISBN: 978-1-53618-615-4
Editor: Grégoire Roux

Chapter 1

Study on Pathogenesity of *Enterobacter* Species: Its Mechanism and Genes/Virulence Factors

Nazish Mazhar Ali, PhD and Iram Liaqat, PhD
Department of Zoology, GC University, Lahore, Pakistan

Abstract

Enterobacter (genus) relatively lesser known member of family enterobacteriaceae, yet with at least fifteen species more or less involved in different human infections. A very little knowledge is available about the pathogenicity and virulence factors of different species of *Enterobacter*. These Gram-negative bacteria possess endotoxins and have complex pathogenetic mechanism as sustain quite a number of putative virulence factors. But still their role in infections in host is not well studied. Some studies showed that *in vitro, Enterobacter* spp. produce some virulence factors after adhesion to host epithelial cells. These factors include enterotoxins, α-hemolysin and thiol-activated cytotoxins (pore-forming). In several Gram-negative bacteria including *Enterobacter* spp., the type III secretion system (TTSS) has proved to be very important with various proteins. This system delivers virulence factors (toxins) in host cells. Thus TTSS plays an important role in pathogen-host interactions. The presence of the TTSS can be used as a general indicator of bacterial virulence. *E. cloacae* a well known pathogenic species mostly isolated from clinical samples, has 27% TTSS genes, a clear indication that this bacteria can damage epithelial cells and phagocytes within the host. *E. cloacae* pathogenesis is majorly contributed by cytotoxic enterotoxins produced by 2-mercaptoethanol and the TTSS. *E. cloacae* strains may induce apoptosis of HEp-2 cells in host as primary strategy to spread systemic infection. Another important factor is the biofilm formation by producing proteinaceous extracellular fibers called curli fimbriae. These fimbriae help bacteria to bind with a variety of host proteins and also adhesion and invasion. These curli fimbriae are encoded by curli gene in *E. cloacae* and other species of *Enterobacter*. In this chapter the virulence factors involved in pathogenesis and the genes associated with these factors in *Entrobacter* spp. shall be discussed.

Keywords: *Enterobacter* spp., pathogenesis, virulence factors, TTSS, biofilm formation, *E. cloacae*

INTRODUCTION

1. *Enterobacter*: Member of Family Enterobacteriaceae

The bacterium *Enterobacter* sp. belongs to family Enterobacteriaceae. The members of Enterobacteriaceae are primarily inhabitants of the lower part of gut in humans and other animals. *Enterobacter* spp. are found in soil and water like *E. cloacae*. *E. aerogenes* can inhabit the intestines of humans and animals and can also be found in sewage. *E. aerogenes* has also been found in dairy products. *Enterobacter* spp. are facultative anaerobes and are Gram-negative bacilli. The size normally ranges 0.6-1 μm in diameter and 1.2-3 μm in length. These bacteria are motile and use peritrichous flagella with class 1 fimbriae. These produce acid via fermenting glucose and show negative methyl red test and positive Voges-Proskauer test. This bacterium shows optimum growth at 30 °C. 80% of these bacteria are encapsulated [1-4].

1.1. Occurrence of Enterobacter spp.

All *Enterobacter* species are found in sewage, soil, water and vegetables. *Enterobacter cloacae* is most frequently isolated species. *Enterobacter* species are also found in humans and animals [5]. The role of this bacterium as an enteric pathogen is not complete it known. However, it is an opportunistic pathogen and has its increasing importance especially in hospital-acquired infections (intensive care units, emergency units, urology units). *Enterobacter cloacae* is less susceptible to chlorination than other bacteria e.g., *Escherichia coli*. It may also be isolated from meat and also present as commensal on the skin of man [6, 7]. *Enterobacter sakazakii* has no importance as an enteric pathogen, but it is a rare cause of meningitis and sepsis in neonates. This bacterium is present environment as well as in food. *Enterobacter agglomerans* is occasionally isolated from clinical specimens but has no importance as an enteric pathogen. *Enterobacter aerogenes*, phenotypically very similar to *E. gergoviae*, is found in dairy products and water [8-10].

2. Infections (Pathogenicity)

Enterobacter spp., especially E. aerogenes and E. cloacae, are associated with nosocomial infections and are spread through feco-oral pathways. These bacteria are considered opportunistic pathogens of animals including humans. *Enterobacter* spp. Infections include cerebral abscess, pneumonia, septicemia, wound, urinary tract (particularly catheter-related UTI), abdominal cavity or intestinal infections and meningitis. Moreover, *Enterobacter* spp. are involved in intravascular device-related infections and also surgical site infections. Some species cause extra-intestinal infections, like, *Enterobacter sakazakii*, causes brain abscesses in infants and with meningitis. A mortality rate for bacterial meningitis is moderate to high (range from 40% -80%) [11-14].

Table 1. Infections caused by *Enterobacter* spp.

Bacteria	Infections	symptoms
Enterobacter spp.	1. Bacteremia 2. lower respiratory tract infections 3. skin and soft-tissue infections 4. urinary tract infections (UTIs) 5. endocarditis 6. intra-abdominal infections 7. septic arthritis 8. osteomyelitis 9. CNS infections 10. ophthalmic infections	*Enterobacter* infections do not have a clinical or distinct presentation that is specific enough to differentiate these infections from other acute bacterial infections.

2.1. Epidemiolog

Majority of *Enterobacter* infections are because of *E. cloacae* and *E. aerogenes* (65-75% and 15-25%, respectively). Different species of this bacterium are found in ICUs (intensive care units) and are involved in 8.6% of nosocomial infections as reported by US Centers for Disease Control and Prevention (CDC). These bacteria are also responsible for bacteremia cases in pediatric hospitals (14%) and 1.5-6% of bacteremia cases in adults. There was an epidemic of septicemia in 1970 which caused by contaminated IV products that ultimately affected nearly 400 patients across the United States. Most epidemiologic aspects of *Enterobacter* infections show the infection rather than the intrinsic virulence of the organism involved. *E. sakazakii* and *E. agglomerans* cause infections which *are* less common than infections caused by *E. cloacae* and *E. aerogenes* [15- 17]. This reflects the probability of special conditions in which infections by the first two species are seen but not the last two species. Infections by *E. sakazakii* are seen in neonates mostly but has been isolated from powdered milk and infant formula milk [18, 19]. Infections caused by *E. agglomerans* are associated with an external or exogenous source. This bacteria grows at 4 °C and is often associated with plants, and can be isolated from cotton. It is predominantly associated with outbreaks due to intravenous solutions contaminations and also stored blood products contamination as well as "cotton fever" in intravenous drug abusers [20-21], *E. agglomerans* although ubiquitous in nature, is not frequent a cause of endogenous nosocomial infections like *E. cloacae* or *E. aerogenes* [22-24]. This probably shows the greater intrinsic susceptibility of *E. agglomerans* than other *Enterobacter* spp. to þ-lactam antibiotics.

2.2. Infectious Dose of Enterobacter spp.

Infectious dose or *Enterobacter* spp. number in an individual is still not well known [25]. However, usually 1000 *Enterobacter* cells are considered infectious. This dose is similar to the infectious dose of other pathogenic bacteria like *Neisseria meningitis*, *Escherichia coli O157* and *Listeria monocytogenes*.

2.3. Mode of Transmission of Enterobacter spp.

Like other members of Enterobacteriaceae family, this bacterium can also be transmitted through the fecal-oral route. Other ways of transmission include direct or indirect contact of mucosal surfaces with infectious agent e.g., these bacteria can be transferred from contaminated hands or contaminated urinals in neonatal units or through transfer to susceptible body sites [26-27].

3. Antibiotic Resistance of *Enterobacter* spp.

Enterobacter species do have an intrinsic resistance to antibiotics like ampicillin and cephalosporins. This resistance enables the bacteria to survive and grow in the intestines of individuals using these antibiotics. Resistance of *Enterobacter* spp. to major antimicrobial agents widely varies. For the þ-lactam antibiotics, the percentage resistance is 8 to 50% for ticarcillin, 9 to 53% for mezlocillin, 7 to 54% for piperacillin, 5 to 63% for cefotaxime, 6 to 59% for ceftazidime, 7 to 44% for aztreonam, 0.3 to 9% for cefepime, and 0 to 4% for imipenem [28-31]. These percentages show the greater activity of broad-spectrum cephalosporins and carbapenems against *Enterobacter* spp. For the aminoglycosides, the percentage resistance to gentamicin is 0 to 51%, the percentage resistance to tobramycin is 0 to 43%, and percentage resistance to amikacin is 0 to 34%. For ciprofloxacin, resistance varies from 0 to 36%, and for trimethoprim-sufamethoxazole, resistance is 0 to 60% of strains [32-34]. These broad ranges of resistance suggest that several factors affect the occurrence of antimicrobial resistance among species of *Enterobacter*.

4. Virulence Factors/Genes and Mechanism Involved

4.1. Toxins

Individuals with more intestinal *Enterobacter* infections are also infected with their own intestinal flora [35]. The virulence factors associated with *Enterobacter* spp. have not been extensively studied till yet because of the relatively low importance as human pathogens, [36, 37]. *Enterobacter* spp. strains produce different toxins like enterotoxins, α-hemolysin and thiol-activated pore-forming cytotoxins similar to Shiga-like toxin II. The virulence factors are delivered by type III secretion system, usually the toxins of various types into the host cells. Type III secretion system genes are also found in other members of enterobacteriaceae e.g., *E.coli* [38], [39-42].

4.2. Types of Toxins

Gram-negative bacteria contain a lipopolysaccharide (LPS) layer. Glycolipid moiety (lipid A) of the LPS, is mainly responsible for endotoxic activity. Endotoxins are rather released when cells are disrupted than being secreted from bacterial cell. Some species of *Enterobacter* are enterotoxigenic, producing shiga-like toxin II-related cytotoxins [43, 44]. An outer membrane protein 32kD (kiloDalton) to be present in 79% of bacterial isolates from cerebral spinal fluid (CSF) and only 9% of non-CSF isolates. Although it is unclear what role this protein plays in pathogenesis, it may be a neurovirulence factor. *E. cloaceae*, is one of the most frequently isolated and clinically important *Enterobacter* species, and is studied for its virulence-associated factors. Some *E. cloaceae* have low-molecular weight, heat-stable, and heat-labile enterotoxin which is sufficient to cause diarrhea [45, 46]. Recently, *E. cloacae* are found to have a Shiga-like toxin II (SLT-II) and are said to be associated with a case of hemolytic-uremic syndrome (HUS). OmpX (18.6 kD), has a direct effect on the ability of *E. cloacae* to invade epithelial tissue. Some species of *Enterobacter* possess serum resistance, while others produce aerobactin (siderophore), and yet others exhibit mannose-sensitive hem-agglutination mechanism . Some species of *Enterobacter* infect hepatitis 2 (Hep-2) cells and become internalized (low numbers of associated bacteria were internalized).

Aerobactin another virulence factor may be an important in establishment of infection as it aids bacterial translocation from the intestine to other tissues and also multiplication of the bacteria in the tissues [47]. *E. cloacae* are usually commensals, they overgrow when selected for by antibiotic treatment as they have the ability to adhere and invade cells. These bacteria can become systemic and can swap genetic material with other bacteria, like *E. coli*. These are serum-resistant and can chelate iron which enables them to survive and spread in the host. Although a little is known about potential virulence characters of these bacteria, still their capsule, which may contribute to serum resistance and resistance to phagocytosis. *E. cloacae* frequently produce aerobactin, which adhere to tissue culture cells, and exhibit mannose-sensitive hem-agglutination, which is possibly result of type 1 fimbriae expression. *E. sakazakii* can adhere to and infect brain microvascular endothelial cells. This is a trait that it shares with others which produce K1 capsule and cause neonatal meningitis [48]. Species belonging to the genus *Enterobacter* are notorious for antimicrobial resistance, earning them acronymal notoriety. *E. cloacae, E. aerogenes,* and most strains of *E. sakazakii* are intrinsically resistant to ampicillin and first- and second-generation cephalosporins as a result of an inducible *ampC* chromosomal β-lactamase that is controlled by both positive and negative regulators [49, 50].

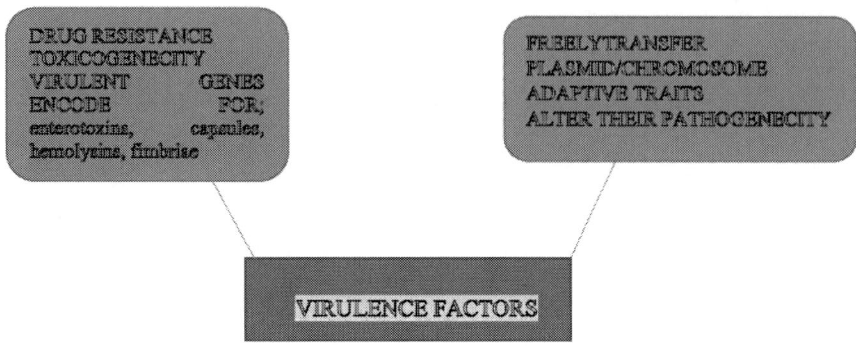

Figure 1. Virulence factors in *Enterobacter* spp.

5. Future Perspective about *Enterobacter* spp. Research

Enterobacter spp. are well adapted for survival and proliferation as the turn of the century approaches. Control of these organisms are quite limited but concise infection control protocols and keen attention to principles of antisepsis can reduce the occurrence of outbreaks. The usual infection control procedures are mostly failing to affect the overall incidence of nosocomial infections by *Enterobacter* spp. Epidemiologic studies should be designed to identify risk factors for the acquisition of *Enterobacter* spp. in the com- munity. The potential role of new broad-spectrum oral cephalosporins in providing a specific pressure which favours *Enterobacter* spp. must be assessed promptly. It is important to give or explain the association between the use of parenteral cephalosporins and *Enterobacter* infections in the hospitals.

Many fundamental questions remain unanswered till yet. Like, what pathogenetic mechanism(s) sets *Enterobacter* spp. apart or distinct clinically from other Gram-negative enteric rods? What favors the occurrence and transmission of these bacteria in solutions and on surfaces of catheters any other medical devices? Why *Enterobacter* spp. show extremely high efficiency as compared to other bacteria in producing disease following infusion or transplantation into uninfected patients? What are the implications of the recently recognized high rate of coin- fection by other pathogens for the diagnostic laboratory and for selection of empiric therapy? What are the main mechanisms that favour the resistance of *Enterobacter* spp. to "fourth- generation" cephalosporins, carbapenems, and fluoroquinolones? What are virulence factors and genes and mechanism of virulence in different species of *Enterobacter*. Can further emergence of resistance be minimized in future? Only with additional basic biomedical research with innovative approaches can give solutions. Further studies can be designed for prevention and treatment of *Enterobacter* spp. infections.

Conflicts of Interest

Authors declare no conflicts of interests.

References

[1] Falkiner, F. R. (1992). *Enterobacter* in hospital. *J. Hosp. Infect.* 20:137–140.

[2] Farmer, J. J., III. (1995). Enterobacteriaceae: introduction and identification, p. 438–449. In P. R. Murray, E. J. Baron, M. A. Pfaller, F. C. Tenover, and R. H. Yolken (ed.), *Manual of clinical microbiology*, 6th ed. American Society for Microbiology, Washington, D.C.

[3] Farmer, J. J., III, M. A. Asbury, F. W. Hickman, D. J. Brenner, and the E. S. Group. (1980). *Enterobacter sakazakii*: a new species of "Enterobacteriaceae" isolated from clinical specimens. *Int. J. Syst. Bacteriol.* 30:569–584.

[4] Farmer, J. J., III, B. R. Davis, F. W. Hickman-Brenner, A. McWhorter, G. P. Huntley-Carter, M. A. Asbuty, C. Riddle, H. G. Wathen-Grady, C. Elias, G. R. Fanning, A. G. Steigerwalt, C. M. O'Hara, G. K. Morris, P. B. Smith, and D. J. Brenner. (1985). Biochemical identification of new species and biogroups of Enterobacteriaceae isolated from clinical specimens. *J. Clin. Microbiol.* 21:46–76.

[5] Ferguson, R., C. Feeney, and V. A. Chirurgi. (1993). *Enterobacter agglomerans*-associated cotton fever. *Arch. Intern. Med.* 153:2381–2382.

[6] Ferret, D., M. Bues-Charbit, C. Granthil, and G. Balansard. (1992). Pharmacoeconomic analysis of antibiotherapy in intensive care. *J. Pharm. Clin.* 11:177–182. (In French).

[7] Finegold, S. M., and C. C. Johnson. (1985). Lower respiratory tract infection. *Am. J. Med.* 79(Suppl. 5B):73–77.

[8] Flaherty, J. P., S. Garcia-Houchins, R. Chudy, and P. M. Arnow. (1993). An outbreak of gram-negative bacteremia traced to contaminated o-rings in reprocessed dialyzers. *Ann. Intern. Med.* 119:1072–1078.

[9] Flynn, D. M., R. A. Weinstein, C. Nathan, M. A. Gaston, and S. A. Kabins. (1987). Patients' endogenous flora as the source of "nosocomial" *Enterobacter* in cardiac surgery. *J. Infect. Dis.* 156:363–368.

[10] Fryklund, B. A., K. Tullus, and L. G. Burman. (1994). Association between climate and *Enterobacter* colonization in Swedish neonatal units. *Infect. Control Hosp. Epidemiol.* 14:579–582.

[11] Al Ansari, N., E. B. McNamara, R. J. Cunney, M. A. Flynn, and E. G. Smyth. (1994). Experience with *Enterobacter bactaeremia* in a Dublin teaching hospital. *J. Hosp. Infect.* 27:69–72.

[12] Almugeiren, M. M., and S. M. H. Qadri. (1991). Etiology of childhood urinary tract infections and antimicrobial susceptibility of uropathogens at a teaching hospital in Saudi Arabia. *Curr. Ther. Res.* 50:454–459.

[13] Andersen, B. M., S. M. Almdahl, D. Sorlie, R. Hotvedt, J. Backer-Chris- tensen, R. B. Nicolaysen, and O. I. Solem. (1990).

Enterobacter cloacaeinfections at the University Hospital in Tromsa. *Tidsskr. Nor. Laegeforen.* 110:342–347. (In Norwegian.)

[14] Kronish, J. W., and W. M. McLeish. (1991). Eyelid necrosis and periorbital necrotizing fasciitis—report of a case and review of the literature. *Ophthalmology* 98:92–98.

[15] Kuhn, I., B. Ayling-Smith, K. Tullus, and L. G. Burman. (1993). The use of colonization rate and epidemic index as tools to illustrate the epidemiology of faecal Enterobacteriaceae strains in Swedish neonatal wards. *J. Hosp. Infect.* 23:287–297.

[16] Anderson, E. L., and J. P. Hieber. (1983). An outbreak of gentamicin-resistant *Enterobacter cloacae* infections in a pediatric intensive care unit. *Infect. Control* 4:148–152.

[17] Ara´ngiz, A. F., R. Alonso, K. Colom, A. Morla, E. Suinaga, and R. Cisterna. (1994). Multicenter study of cefotaxime resistance—1993. *Rev. Esp. Quimioterap.* 7:57–61. (In Spanish.)

[18] Aubert, G., P. P. Levy, A. Ros, R. Meley, B. Meley, A. Bourge, and G. Dorche. (1992). Changes in the sensitivity of urinary pathogens to quinolones between 1987 and 1990 in France. *Eur. J. Clin. Microbiol. Infect. Dis.* 11:475–477.

[19] Ballow, C. H., and J. J. Schentag. (1992). Trends in antibiotic utilization and bacterial resistance. Report of the National Nosocomial Resistance Surveillance Group. *Diagn. Micobiol. Infect. Dis.* 15:37S–42S.

[20] Bamberger, D. M., and S. L. Dahl. (1992). Impact of voluntary vs enforced compliance of third-generation cephalosporin use in a teaching hospital. *Arch. Intern. Med.* 152:554–557.

[21] Baquero, F. (1995). Pneumococcal resistance to þ-lactam antibiotics: a global geographic overview. *Microb. Drug Resist.* 1:115–120.

[22] Bauernfeind, A., R. Jungwirth, and S. Schweighart. (1989). In-vitro activity of meropenem imipenem, the penem HRE 664 and ceftazidime against clinical isolates from West Germany. *J. Antimicrob. Chemother.* 24(Suppl. A):73–84.

[23] Bellon, J., and R. P. Mouton. (1992). Distribution of beta-lactamases in Enterobacteriaceae: Indoor versus outdoor strains. *Chemotherapy* 38:77– 81.

[24] Bittner, M. J., D. L. Dworzack, L. C. Preheim, R. W. Tofte, and K. B. Crossley. (1983). Ceftriaxone therapy of serious bacterial infections in adults. Antimicrob. *Agents Chemother.* 23:261–266.

[25] Bonadio, W. A., D. Margolis, and M. Tovar. (1991). *Enterobacter cloacae* bacteremia in children: a review of 30 cases in 12 years. *Clin. Pediatr.* 30:310–313.

[26] Bone, R. C. (1993). Gram-negative sepsis: a dilemma of modern medicine. *Clin. Microbiol. Rev.* 6:57–68.

[27] Bonfiglio, G., S. Stefani, and G. Nicoletti. (1994). In vitro activity of cefpirome against beta-lactamase-inducible and stably derepressed Enterobacteriaceae. *Chemotherapy* 40:311–316.

[28] Bouza, E., M. G. de la Torre, E. L. A. Erice, J. M. Diaz-Borrego, and L. Buzo′n. (1985). *Enterobacter bacteremia*—an analysis of 50 episodes. *Arch. Intern. Med.* 145:1024–1027.

[29] Breyer, S., S. M. Feistauer, H. Burgmann, M. Georgopoulos, and A. Georgopoulos. (1991). Epidemiology and spectrum of causative organisms of urinary tract infections. *Themenheft Harnwegsinfekt.* 23/24:533–536. (In German.)

[30] Buchholz, D. H., V. M. Young, N. R. Friedman, J. A. Reilly, and J. M. R. Mardiney. (1971). Bacterial proliferation in platelet products stored at room temperature: transfusion-induced *Enterobacter sepsis*. *N. Engl. J. Med.* 285:429-433.

[31] Burchard, K. W., D. T. Barrall, M. Reed, and G. J. Slotman. (1986). *Enterobacter bacteremia* in surgical patients. *Surgery* 100:857–861.

[32] Burwen, D. R., S. N. Banerjee, R. P. Gaynes, and the N. N. I. S. System. (1994). Ceftazidime resistance among selected nosocomial gram-negative bacilli in the United States. *J. Infect. Dis.* 170:1622–1625.

[33] Bush, K., G. A. Jacoby, and A. A. Medeiros. (1995). A functional classification scheme for þ-lactamases and its correlation with molecular structure. *Antimicrob. Agents Chemother.* 39:1211–1233.

[34] Carsenti-Etesse, H., J. Durant, F. D. Salvador, M. Bensoussan, F. Ben- soussan, C. Pradier, A. Thabaut, and P. Dellamonica. (1995). In vitro development of resistance of *Streptococcus pneumoniae* to þ-lactam antibiotics. *Microb. Drug Resist.* 1:85–94.

[35] Chamberland, S., J. L'Ecuyer, C. Lessard, M. Bernier, P. Provencher, M. G. Bergeron, and the C. S. Group. (1992). Antibiotic susceptibility profiles of 941 Gram-negative bacteria isolated from septicemia patients throughout Canada. *Clin. Infect. Dis.* 15:615–628.

[36] Chenoweth, C., and D. Schaberg. (1990). The epidemiology of enterococci. *Eur. J. Clin. Microbiol. Infect. Dis.* 9:80–89.

[37] Chow, J. W., V. L. Yu, and D. M. Shlaes. (1994). Epidemiologic perspectives on *Enterobacter* for the infection control professional. *Am. J. Infect. Control* 22:195–201.

[38] Chung, K. I., T. H. Lim, Y. S. Koh, J. H. Song, W. S. Kim, J. M. Choi, and Y. H. Auh. (1992). Nosocomial pneumonia in medico-surgical intensive care unit. *J. Korean Med. Sci.* 7:241–251.

[39] Cohen, S. P., W. Yan, and S. B. Levy.(1993). A multidrug resistance regulatory chromosomal locus is widespread among enteric bacteria. *J. Infect. Dis.* 168:484–488.

[40] deChamps, C., M. P. Sauvant, C. Chanal, D. Sirot, N. Gazuy, R. Malhuret, J. C. Baguet, and J. Sirot. (1989). Prospective survey of colonization and infection caused by expanded spectrum-þ-lactamase-producing members of the family Enterobacteriaceae in an intensive care unit. *J. Antimicrob. Chemother.* 27:2887–2890.

[41] de Oliviera, G. F., L. Barrucand, C. M. N. David, and P. P. G. Filho. (1990). The use of new generation beta-lactams and bacterial resistance in the Hospital Universitario da UFRJ. *Folha Med.* 101:237–242. (In Portuguese.)

[42] Deusch, E., A. End, M. Grimm, W. Graninger, W. Kleptko, and E. Wolner. (1993). Early bacterial infections in lung transplant recipients. *Chest* 104: 1412–1416.

[43] Hopkins, J. M., and K. J. Towner. (1990). Enhanced resistance to cefotaxime and imipenem associated with outer membrane protein

alterations in *Enterobacter aerogenes. J. Antimicrob. Chemother.* 25:49–55.

[44] Lee, E.-H., M. H. Nicholas, M. D. Kitzkis, G. Pialou, E. Collatz, and L. Gutmann. (1991). Association of two resistance mechanisms in a clinical isolate of *Enterobacter cloacae* with high-level resistance to imipenem. *Antimicrob. Agents Chemother.* 35:1093–1098.

[45] Jones, R. N., M. E. Erwin, M. S. Barrett, D. M. Johnson, and B. M. Briggs. (1991). Antimicrobial activity of E-1040, a novel thiadiazolyl cephalosporin, compared with other parenteral cephems. *Diagn. Microbiol. Infect. Dis.* 14:301–309.

[46] Jones, R. N., M. A. Pfaller, P. C. Fuchs, K. Aldridge, S. D. Allen, and E. H. Gerlach. (1989). Piperacillin/Tazobactam (YTR 830) combination—comparative antimicrobial activity against 5889 recent aerobic clinical isolates and 60 Bacteroides fragilis group strains. *Diagn. Microbiol. Infect. Dis.* 12:489– 494.

[47] Kappstein, I., G. Schulgen, T. Friedrich, P. Hellinger, A. Benzing, K. Geiger, and F. D. Daschner. (1991). Incidence of pneumonia in mechanically ventilated patients treated with sucralfate or cimetidine as prophylaxis for stress bleeding: bacterial colonization of the stomach. *Am. J. Med.* 91(Suppl. 2A):125S–131S.

[48] Karnad, A., S. Alvarez, and S. L. Berk. (1987). *Enterobacter pneumonia. South. Med. J.* 80:601–604.

[49] King, A., C. Boothman, and I. Phillips. (1990). Comparative in vitro activity of cefpirome and cefepime, two new cephalosporins. Eur. J. Clin. *Microbiol. Infect. Dis.* 9:677–685.

[50] Weber, D. A., and C. C. Sanders. (1990). Diverse potential of þ-lactamase inhibitors to induce class 1 enzymes. *Antimicrob. Agents Chemother.* 34: 156–158.

In: Recent Developments in *Enterobacter* … ISBN: 978-1-53618-615-4
Editor: Grégoire Roux

Chapter 2

EFFICACY OF NANOPARTICLES FROM BIOLOGICAL SOURCES OFFERS NOVEL APPROACH TO CONTROL PATHOGENIC BACTERIA OF FAMILY ENTEROBACTERIACEAE

Sumaira Mazhar[1], PhD, Beenish Sarfaraz[1], Iram Liaqat[2,*], PhD and Aisha Waheed Qurashi[1,†], PhD

[1]Department of Biology, Lahore Garrison University, Sector C Phase 6 DHA Lahore, Pakistan

[2]Microbiology Laboratory, Department of Zoology, GC University Lahore Pakistan

ABSTRACT

Introduction: Members of family Enterobacteriaceae are pathogenic and offer potential health risks to human and animals. These members

---

* Corresponding Author's Emails: iramliaq@hotmail.com; dr.iramliaqat@gcu.edu.pk.

† Corresponding Author's Emails: aieshawaheed@yahoo.com; aishawaheedqureshi@lgu.edu.pk.

form pathogenesis by using different biochemical and molecular strategies to cope with host defense system. The use of classic approaches like antibiotics besides being expensive to implement poses serious threats to ecosystems. The efficacy of reported use of different nanoparticles as antimicrobial agent is one of the novels and yet to explore strategies for control of pathogenic microorganisms from family Enterobacteriaceae. Conventionally physical and chemical methods are used for the synthesis of diverse types of nanoparticles. Although these methods are giving promising results but there are some limitations regarding the use of methods as they are not ecofriendly and use toxic chemicals. Therefore, interest in the synthesis of nanoparticles by using biological sources has been amplified. Biological methods used for the synthesis of nanoparticles are not only cost effective, can be easily scaled up but also environmentally friendly. Biological sources like plant and microbes including bacteria, algae, and fungi are well known for their potential to produce diverse range of nanoparticles.

Conclusion: Current review will highlight the plants and microbes having antimicrobial potential for using in the synthesis of nanoparticles for the control of pathogenic Enterobacteriaceae in safe way.

Keywords: algae, antibacterial activity, biological methods, enterobacteriaceae, fungi, nanoparticles, pathogens, plants

1. INTRODUCTION

The family Enterobacteriaceae has been reported to be the most heterogeneous and largest collection of medically important gram negative bacilli including 27 genera. Some members being common gut flora becomes opportunistic pathogens like *Escherichia coli, Klebsiella pneumoniae, Proteus vulgris*. However, members like *Salmonella, Shigella, Yersinia pestis* are the human pathogenic members of this family. Enterobacteriaceae has been the most notorious with reference to nosocomial infections and becoming matter of serious concern with reference to antibiotic resistance development [1]. The first most notorious case of this family was reported in 1976. Among multiple nosocomial infections reported so far, the most common are Urinary tract infections, meningitis and lungs related infections [1, 2]. Where an outbreak of septicemia was reported with the use of infected intravenous solution source [3]. Glucose composition-based fluids were found to favor the replication of the bacterial growth of these members. The prevalence of nosocomial infections due to Enterobacteriaceae becomes more pronounced due to their survival on dry surfaces. The other reported examples of these infections are from use of humidifiers and contaminated instruments and related units [4, 5]. The main reason for the pathogenesis of Enterobacteriaceae members is presence of their specific traits like adhesions, endotoxin, and siderophores to acquire iron [6]. These traits are involved in resisting the pathogens of family Enterobacteriaceae against antibiotics. For example, Enteropathogenic *Escherichia coli* (EPEC) has been reported to have adherence factors for attachment with mucosal lining of bowel and ultimately results in watery diarrhea among infants. Production of plasmid encoded heat-labile exotoxin is another trait in members of this family. Similarly, *Klebsiella pneumoniae* cause mainly respiratory and *K. oxytoca* causes urinary tract infections for having large polysaccharide capsules. Infections by family Enterobacteriaceae becomes notorious due to their resistance against antibiotics which are considered to be effective against any disease, however, its efficacy is drastically reduced when bacteria develop strong resistance against these antibiotics and there is need to develop specific measures to overcome this resistance [7]. The utilization of Nanoparticles seems promising approach to combat this antibiotic resistance problem. Development of biofilms is another important trait observed as defensive strategy within bacteria to cope the effect of antibiotic resistance. Within these communities' bacteria entrap themselves in the form of aggregates within a sheath of polysaccharides in non-motile state. The development of microbial biofilm is a great health concern due to more pronounced antibiotic resistance development in bacteria [8]. Such approaches that involve the use of NPs in the form of coatings for different devices either the implantable devices diagnostics or as vaccines as antibacterial agents. While adopting the mechanisms of NPs induction of oxidative stress as well as non-oxidative mechanisms seems promising as antibacterial agents. The efficacy of nanoparticles is more enhanced as compared to antibiotics because to show bacteriostatic effect Nanoparticles do not need to penetrate within the cells. It sparks the interest of scientists that it will show less resistance in bacteria as compared to antibiotics [9]. The efficacies of nanoparticles have been reported against diverse species of bacteria including gram positive and negative bacterial species. NPs in particular have demonstrated broad-spectrum antibacterial properties against both Gram-positive and Gram-negative bacteria. Wang et al. [10] reported the growth inhibition role of ZnO NPs against *Staphylococcus aureus* while silver nanoparticles showed inhibiting effect against *E. coli* and *Pseudomonas aeruginosa*. Evidence of growth inhibiting efficacy of Nanoparticles against catheters biofilms also exist [11, 12]. Similarly, Nano-TiO_2 based implants are widely used in dental and orthopedic devices for their effectivity as antimicrobial agents against biofilm formation. Nanoparticles based coating in the form of coatings on significantly reducing the growth of *S. aureus* as reported by Wang et al., [10]. Nanoparticles show antibacterial effects by following methods either the disruption of cell membranes, and interacting with DNA and Proteins and production of reactive oxygen species in the cells to kill bacterial cells. For the control of ever-growing disease resistance in Enterobacteriaceae, the developments of new and innovative approaches to control the disease infection are significant for research. This chapter describes the overview of research work related to plant based nanoparticles and their efficacy in control of infection.

With the unique properties of Nanomaterial and ability to react with biomolecules, Nanoparticles have been in great demand from last twenty years [13-14]. However, synthesis of metallic nanoparticles besides being costly requires toxic chemicals [15-16]. Therefore, utilization of green synthesis approaches have been of considerable importance in last few years using phytochemicals, yeast /fungal sources and algae [17]. Plant metabolites convert metallic ions to metallic nanoparticles [18]. Plants metabolites play important role to synthesize nanoparticles being ecofriendly as in a single step, without toxicants desired products are obtained [19].

Being the major contributing component of secondary metabolites, plants have great significance. These secondary metabolites have diverse applications in agriculture, Pharmaceutical, clinical and food industries.

One of such applications includes utilization of plants for nanoparticle synthesis. Different parts of plants and different species of plants are used for the synthesis of nanoparticles. With improvement in life style of human being on earth, utilization of nanoparticles as the technological approach increases that renders the ecosystem at miserable situation. This becomes more pronounced when applications are very diverse. For examples their applications include different cosmetic products, utilization in drugs etc. For the proper utilization of this technology green synthesis of nanoparticles is more in trend to meet the demands of sustainability of environment.

Current chapter deals with the summarized overview of Green synthesis of significantly four nanoparticles (silver, gold, copper and iron) having biomedical applications in the last decade. Biologically derived molecules play role as bio reductant for the synthesis of aforementioned four nanoparticles from metallic oxides compounds. The physiochemical features associated with the nanoparticles that influence their antibacterial activity has also been reviewed. This chapter is a humble effort to compile the last decade research on the biological synthesis of nanoparticles and efficacy of these NPs to control the pathogenic members of family Enterobacteriaceae.

2. BIONANOPARTICLES

2.1. Plant Nanoparticles

2.1.1. Plant Silver Nanoparticles

Silver utensils have been in use for their effective antimicrobial activities. At reduced particle size to nanometer range, bactericidal effect of silver has been found to be increased. Moreover, the antibacterial effect of silver based nanoparticles showed more pronounced effects against different members of Enterobacteriaceae e.g., *Streptococcus pyogens, E. coli etc.* and other pathogenic microbes [20]. Synergistically with combination of other metallic nanoparticles or antibiotics, efficacy of silver nanoparticles have found to show promising effect against various Gram-negative and Gram positive bacteria [21]. Biological synthesis of silver nanoparticles (Ag-NPs) using plant metabolites Table 1 help to reduce silver metal into silver nanoparticles. Makarov et al. [22] reported different secondary metabolites of plants like sugars, phenolic compounds and their derivatives, proteins and terpenoids as bio reducing and bio stabilizing agents to convert silver into silver nanoparticles. Diverse efficacy and applications of silver nanoparticles either in water cleanliness, medical implants and having other antimicrobial activities [22] need of biogenic synthesis of silver nanoparticles increases.

Table 1. Production of nanoparticles by plants

Nanoparticles	Plant	Shape	Size (nm)	Source	References
Silver	*Pelargonium graveolens*	Quasilinear super	16–40	$AgNO_3$	[264]
	Carica papaya	Spherical	50–250	$AgNO_3$	[200]
	Jatropha curcas (latex)	Face-centered cubic	10–20	$AgNO_3$	[265]
	Mentha Piperita (leaf extract)	Spherical	5–30	$AgNO_3$	[266]
	Hibiscus Rosa sinensis	Spherical	14	$AgNO_3$	[267]
	Solanum torvum	Spherical	14	$AgNO_3$	[268]
	Sesuvium portulacastrum	Spherical	5 to 20	$AgNO_3$	[269]
	Acalypha indica	Spherical	10 –26	$AgNO_3$	[270]
	Chenopodium albumleaves	Spherical, Triangular	10–30	$AgNO_3$	[271]
	Avicenna marina	Spherical	71–110	$AgNO_3$	[201]
	Arbutus unedo	Spherical	30	$AgNO_3$	[272]
	Euphorbia prostrata	Rod	25-80	$AgNO_3$	[273]
	Vitis vinifera	Spherical	30-40	$AgNO_3$	[274]
	Rumex hymenosepalus	face-centered cubic, Hexagonal	2 to40	$AgNO_3$	[275]
	Bridelia retusa	Spherical	52 -96	$AgNO_3$	[276]
	Trianthemadecandra (Aizoaceae)	Spherical	36–74	$AgNO_3$	[277]
	Mentha piperita	Spherical	90	$AgNO_3$	[278]

Nanoparticles	Plant	Shape	Size (nm)	Source	References
Gold	*Tamarindus indica*	Triangular	20-40	$HAuCl_4$	[264]
	Mentha piperita	Spherical	150	$HAuCl_4$	[278]
	Trianthema decandra	Spherical, Cuboil,	33–65	$HAuCl_4$	[40]
	Euphorbia hirta	Spherical	71	$HAuCl_4$)	[204]
	Annona muricate	Spherical	25.5	$HAuCl_4$	[279]
	Scutellaria barbata	Spherical	0.4 μm to 1 μm	$HAuCl_4$	[280]
	Salix alba	--	50-80	HAuC14	[281]
	Solanum nigrum Leaves	Spherical	2.7 - 32.7	HAuC14	[203]
	Ziziphus zizyphus	Spherical, Triangular, Hexagonal	50	$HAuCl_4$	[206]
Iron	*Hordeum vulgare, Rumex acetosa*	Spherical	Up to 30	$FeCl_3.6H_2O$	[18]
	Sorghum spp	Irregular cluster	50 nm	$FeCl_3.6H_2O$	[57]
	Mimosa pudica	Spherical	60–80 nm	$FeSO_4.7H_2O$	[282]
	Caricaya papaya	Irregular	Irregular	$FeCl_3 \cdot 6H_2O$	[57]
	Mangifera indica	Nanorods	15 ± 2	$FeSO_4 \cdot 7H_2O$	[283]
	Salvia officinalis	Spherical	5–25 nm	$FeCl_3$	[284]
	Punica granatum	Spherical	10–30 nm	$FeCl_3 \cdot 6H_2O$	[285]
	Ulex europaeus Pinus pinaster Sapindus mukorossi	Rod-like	50 nm	$Fe(NO_3)_3 \cdot 9H_2O$	[286]
	Eucalyptus sp.	Spherical	80 nm	$Fe(NO_3)_3 \cdot 9H_2O$ $FeCl_3 \cdot 6H_2O$	[287]
	Camellia sinensis (Black tea)	Round	40–50 nm	$FeSO_4 \cdot 7H_2O$	[67]
	Mansoa alliacea	-	18 nm	$FeSO_4 \cdot 7H_2O$	[288]
	Andean blackberry	Spherical	40–70 nm	$FeSO_4 \cdot 7H_2O$	[69]
	Amaranthus spinosus	Spherical	58–530 nm	$FeCl_3$	[289]
	Eucalyptus	Cubic	40–60 nm	$FeCl_3$	[62]
	Camelliasinensis (Green tea)	Spherical	70 nm	$Fe(NO_3)_3$	[18]
	Eucalyptus tereticornis	Spherical	50–80 nm	$FeCl_3$	[290]
	Coffee	Triangular	19.6 ± 25.8 nm	$FeCl_3 \cdot 6H_2O$	[291]

Table 1. (Continued)

Nanoparticles	Plant	Shape	Size (nm)	Source	References
	Camellia sinensis *Syzygium aromaticum* *Mentha spicata*	Spherical	60 nm	$FeCl_3 \cdot 6H_2O$	[292]
	Punica granatum juice *Lawsonia inermis*		21–32 nm	$FeSO_4 \cdot H_2O$	[207]
Copper	*Capparis zeylanica/leave*	Cubical	50 –100	$CuSO_4$	[293]
	S. aromaticum /Flower	Spherical	14 –50	$CuSO_4$	[294]
	O. sanctum Leaves	—	77	$CuSO_4$	[295]
	Datura meta L/.Leaves	Spherical	15 –20	$CuSO_4$	[296]
	Citrus limon /Fruit	Spherical	60 –100	CuCl	[44]
	Papaya/ Leaves	Spherical and crystalline	20	CuCl	[297]
	Citrus medica Linn./Fruit	Crystalline	20	$CuSO_4$	[298]
	Euphorbia esula L/ Leaves	Spherical	20 –110	$CuSO_4$	[299]
	Pistachio /Leaves	Crystallized nanowires	9	CuCl	[300]
	Malvasylvestris/ Leaves	Spherical	14 nm	CuCl	[301]

The use of plants to synthesis Ag-NPs have already been summarized briefly (Table 1). Various plant compounds like polyol and other secondary metabolites result in the reduction of silver ions and finally stabilizing in the form of produced nanoparticles. Being the noble metal nanoparticles, silver nanoparticles are unique for having electrical conductivity, stability I n chemical composition and bactericidal properties [23]. The synthesis procedure for the development of silver nanoparticles involves the washing of collected plant followed by drying either air drying or blender mediated drying. The dried plant material is boiled with deionized distilled water and filtration [24]. Silver nanoparticles have been reported to show more pronounced results when use in combination with poly (vinyl alcohol) and chitosan (CS). Similarly lower concentrations of nano silver were found to be effective against post-surgery-related infections thus increasing their efficacy along with antibiotics [9]. Efficacy of silver nanoparticles as antibacterial agents increased against members of Enterobacteriaceae when used with different antibiotics like cefotaxime, ceftazidime, meropenem, ciprofloxacin and gentamicin. However, low concentrations were most suitable when tested against multidrug resistant bacteria and having no cytotoxic effects [25].

Different techniques (used all at once or used all at a time) to characterize the newly synthesized nanoparticles. The selection of choice of method depends on property to analyze. The techniques used to describe the size, shape and structure of NPs are microscopy techniques like Transmission electron microscopy (TEM), Atomic Force Microscopy (AFM), High-resolution transmission electron microscopy (HRTEM). For analysis of magnetic properties of nanoparticles superconducting quantum interference device magnetometry (SQUID) is used to check the composition of elements, optical properties and other common and more specific physical properties. To analyze the structure of crystal, X-ray diffraction (XRD) is used [26].

2.1.2. Plant Gold Nanoparticles

Biogenic synthesis for the production of gold nanoparticles has been the most promising approach (Table 1). Khan et al. [27] reported the developments in improving the properties of gold nanoparticles. This approach overcomes the negative and hazardous effects of chemical methods for synthesizing gold nanoparticles. Gold being one of the precious metal, also seems effective as antimicrobial agent for having large surface area and thus showing better interaction with microorganisms. For example, gold nanoparticles synthesis from different plant species are found to be proficient in showing shape either spherical or triangular with size (50-100 nm) with efficacy as antimicrobial against different members of family Enterobacteriaceae. Plant metabolites being reducing agents reduces gold metal ions to synthesis gold nanoparticles [24]. However, efforts are switching towards different approaches for increasing the efficacy of gold nanoparticles with increasing antimicrobial properties [9] as optical properties are changed with their shape and sizes [28]. The shape and size can be regulated by variable concentrations [29]. Del Maria et al., [30] reported the plant derived gold nanoparticles from *Annona muricata* exhibiting diverse antimicrobial properties of the nanoparticles diverse microorganisms from Gram-positive and Gram-negative species. With distinct characteristic features, gold nanoparticles have diverse applications especially in fields of Medicines and biology. The efficiency increases when gold nanoparticles are conjugated with other molecules like antibiotics or vaccines. It is important to describe here that evidences for the direct effect of gold NP with no functional role however, their conjugation with biomolecules have been reported to show promising antibacterial effect [31]. Payne et al. [32] reported that minimum inhibitory concentration of gold nanoparticles conjugated with kanamycin was less against members of family Enterobacteriaceae as compared to kanamycin alone. On the other hand results of Rattanata et al. [33] were also similar when using gallic acid and gold conjugates with gallic acid were tested against members of food borne pathogens. In most of the reported literature, generally metals as precursors in the presence of some reducing agents have been in use for synthesis of gold nanoparticles [34]. Similarly as a green technology and its cost effectiveness, utilization of plants besides microorganism being used for the synthesis of gold nanoparticles [35-36]. Green technology has also an edge in the way that plant derived biomolecules act as reducing agents which are far more ecofriendly and preferred as compared to chemical reducing agents [37-39], Au-NPs are also precious in the way that they have promising anticancerous activities e.g., Au-NPs synthesized from plant extracts of *Couroupita guianensis* showed anticancer activity [40]. Many efforts have tested the use of different parts of plants for synthesising of gold NPs be it fruits, leaves seeds, bark and biomolecules.

Different plant leaves for example leaf extract from *Jatropa curcas* L, [41], *Ocimum sanctum*, *Coriandrum sativum* [42] and *Murraya koenigii* [43] have been reported for the synthesis of gold NPs. Fruits involved in synthesis of AuNps includes citrus fruit [44], *Emblica officinalis* [45] pear [46], *Prunus domestica* [47]. Similarly synthesis of gold nanoparticles from bark extracts of *Cassia fistula* [48], *Breynia rhamnoides* stem [49] *Dalbergia sissoo* [50]. Synthesis from seeds involve seeds of *Abelmoschus esculentus* [51]seed oils from *Ricinus communis* [52]. Biomolecules from plants, are peptide macromolecule [53], DNA/proteins [54], DNA and RNA [55] can be used for the synthesis of AuNPs.

2.1.3. Plant Iron Nanoparticles

Biogenic synthesis of Fe-NPs using diverse parts of plants offer attractive, simple and relatively cheap approach for the various applications [56]. Numbers of plants are known to synthesize Fe-NPs using various sources of iron salts as iron precursor. Metabolites released by plants in media transform the iron ions of salts into the iron nanoparticles within few minutes. Metabolites produced by plants (amino acids, simple sugars, polyphenols and phenol derivative compounds, different alkaloids and multiple coenzymes) as a bio reductant and bio stabilizers are involved in the biosynthesis reaction [57]. Culturing conditions (varying iron concentration, volume and composition of plant extract, temperature, incubation period) have significant impact on the synthesis of Fe-NPs [58]. Biosynthesis of Fe-NPs using plant extract is a simple method just involve the preparation of plant extract and its mixing with iron solutions in controlled laboratory conditions. Different parts of plants can be used for the preparation of its extract like leaves, seeds, fruits (Table 1). Plants extract obtained using leaves of *Tridax procumbens* [59], *Castanea sativa*, *Eucalyptus globules* [60], *Dodonaea viscosa* [61], *Eucalyptus tereticornis*, *Melaleuca nesophila* B, *Rosmarinus officinalis* [62] *Punica granatum* [63], *Colocasia esculenta* [64] *Amaranthus dubius* [65], *Ulex europaeus*, *Pinus pinaster* [66] *Camellia sinensis* [35, 67] Fruit extracts of various plants are also reported for synthesis of iron nanomaterials *Terminalia chebula* [68], *Passiflora tripartitavar* [69], *Piper longum* [70]. Seed extract of *Sterculia foetida* [41], *Syzygium cumini* [71] were reported to be use for the synthesis of iron nanoparticles. Analysis of nanoparticles by FTIR showed that plant metabolites when used for synthesis of nanoparticle act as capping agent of INPs [9].

2.1.4. Plant Copper Nanoparticles

Different salts of copper are employed for Cu NPs synthesis which includes cupric acetate (monohydrate), Copper chloride di-hydrate and Copper sulfate pentahydrate. pH, temperature, concentration of chemicals influences properties of these Cu Nanoparticles. Table 1 [72]. Copper oxide nanoparticles (CuO-NPs) were synthesized using a green synthesis approach. Like other nanoparticles various sources of plant extract reported for synthesizing copper nanoparticles. Various salts were of copper were transformed into copper nanoparticles using leaf extract of *Cymbopogon flexuosus* [73], *Ficus carica* (Gultekin et al. [74], *Psidium guajava* [75], *Uncaria gambir* [76]. Extract from *Punica granatum* peel was used as bioreducing agent for copper nanoparticles synthesis. $CuSO_4 \cdot 5H_2O$ solution was combined with the peel extract to obtain NPs that were in the range size of 15-20 nm [77]. Copper nanoparticles (CuNPs) were made using fruit extract of *Rhus coriaria*. The synthesized CuNPs showed their characteristic peak at 560 nm using UV–Visible spectra. CuNPs display an absorption peak at 560 nm [78]. Green tea (*Camellia sinensis*) was used amalgamated with β-cyclodextrin. Green tea is naturally enriched with Polyphenols, which helped to reduce copper ions to Cu-Nps. Absorption peak was observed at 659nm using UV-Visible spectroscopy [79].

Caesalpinia bonducella seed extract was employed to make copper nanoparticles. These nanoparticles were observed for their electrocatalytic properties as well. These seeds are packed with active compounds like citrulline, phytosterinin, β-carotene, and flavonoids which have antioxidant potential as well as they can as capping and stabilizing agents [80]. In another study, *Eclipta prostrata* leaf extract was combined with copper acetate solution for obtaining stable nanoparticles. These NPs were deemed much safer as compared to chemically synthesized NPs with an average size (23-57 nm). These phytosynthesized CuNPs showed considerable cytotoxic activity in HepG2 cell, which could be a promising treatment against cancer [81]. Aqueous extract of *Capparis spinosa* was combined with copper sulfate solution to create copper nanoparticles and furthermore they were assessed for their antibacterial properties again some pathogenic bacterial strains. Adding plant concentrate to the copper sulfate solution, the shade of the solution changed from light blue to yellowish green. Electron microscopy showed the molecule size (17-41 nm [82]. Ginger rhizome based copper nanoparticles (CuNPs) were produced for the reduction of congo red (CR), etc. as a green science approach [83].

2.2. Algal Nanoparticles

2.2.1. Algal Silver Nanoparticles

The syntheses of nanoparticles using biological sources have several advantages over chemical synthesized particles as their sizes can be modified for diverse medical applications [84]. Syntheses of nanoparticles from diverse algal sources considered as safe for environment and cost effective offer attractive opportunity to explore them for its applications in biomedical and industrial areas [85]. Most of studies have reported silver nanoparticles to serve as an efficient barrier for reducing growth of most of the microorganism [86-87]. Silver is a safe and economically cheap source against pathogens over the past centuries. These are non-toxic, cost effective, environmentally friendly, capable of dealing wide range of diseases causing microorganisms and used in the formulation of various medical devices because of its significant potential as antimicrobial agents [88-90]. Algae occupied various habitat has captured a major attention because of its secondary metabolites with huge impact in the medicinal and pharmaceutical fields. Various secondary metabolites with medicinal use isolated from different environmental sources can be considered as an essential tool for nanotechnology [91] (Table 2).

Use of microorganisms for the synthesis of Ag-NPs has recently received remarkable attention for having potential to generate nanoparticles of various shapes and sizes [92]. Small sized primitive microscopic plants like microalgae in comparison to advance higher plants have important advantages for synthesizing NPs, as they propagate very fast [93-94]. Synthesis of these silver NPs using extracts of *Padina tetrastomatica* extract in the formation of reddish-brown color in media after the incubation period of 72 hrs. [95-97]. *Plectonema boryanum,*

Anabaena, *Calothrix, Spirulina platensis* and *Leptolyngbya* has been reported to biosynthesize extracellular Ag-NPs [84, 98, 99]. Algal extract of *Botryococcus braunii* and *Caulerpa serrulata* respectively can be used to synthesise Ag-NPs [100-101]. Their crystalline morphology was confirmed by X-ray Diffraction (XRD) spectrum and the peak record by the Scherrer method was used for the estimation of average size of the crystalline Ag-NPs [100-102]. *Chlamydomonas reinhardtii* and *Chlorococcum humicola* were reported to produce both intracellular and extracellular Ag-NPs [103-04]. *Spirulina platensis*, *Chlorella vulgaris* and *Scenedesmus obliquus* were reported to synthesis Ag-NPs using two different approaches. First approach involved formulation of suspension in 1 mM aqueous $AgNO_3$ solution using thoroughly washed algal biomass and the second approach involved individual culturing of algae in culture media having the $AgNO_3$ solution of same concentration [105]. Polysaccharides and/or proteins present within the microbes may be involved actively in the synthesis of silver nanoparticles as revealed with FTIR spectral analyses. Proteins extracted in the media perform two major dual functions one was the reduction of Ag^+ and second was control of shape in the synthesis nano-Ag particles [106]. NADH-reductase contribute in the synthesis of Ag-NPs by transferring electron from NADH and results in reduction of silver ions to Ag-NPs, the enzyme is oxidized and this process continues [107].

Ulva lactuca, Gelidiella sp., *Sargassum tenerrimum* and *Pithophora oedogonia* have been also reported for synthesising Ag-NPs. Marine algae with rich organic compounds not only used for synthesizing but also coating the NPs. Algal species are considered as rich sources of various secondary metabolites like alkaloids, steroids, phenols, saponins and carbohydrates benefit by reducing and stabilizing NPs [108-111] . Biosynthesis of extracellular Ag-NP observed in *Padina pavonica* was mostly spherical in shape [86]. Terpenoids significantly contribute in biosynthesis of Ag-NPS by oxidation of aldehydic to carboxylic acids [112-113].

Table 2. Production of nanoparticles by algae

Nanoparticles	Algae	Shape	Size (nm)	Source	References
Silver	*Chaetoceros calcitrans, Cuscuta salina, Isochrysis galbana*	Spherical	53-71.9	$AgNO_3$	[223]
	Urospora sp.	Spherical	20 to 30	$AgNO_3$	[224]
	Pithophora oedogonia	Cubical and hexagonal shaped	25–44	$AgNO_3$	[111]
	Padina tetrastomatica	Spherical	33.75	$AgNO_3$	[87]
	Sargassum ilicifolium	Cubic and hexagonal	-	$AgNO_3$	[225]
	Gracilaria birdiae	Spherical	20.-94.9	$AgNO_3$	[302]
	Sargassum tenerrimum	Spherical	20	$AgNO_3$	[110]
	Gelidiella acerosa	Spherical	22	$AgNO_3$	[109]
Gold	*Turbinaria conoides*	Square, Rectangular, cubic and triangular	60	$AuCl_3$	[126]
	Gracilaria Corticata	-	45-57	$HAuCl_4$	[125]
	Padina Tetrastromatica, Turbinaria Ornata	Cubic	18-90	$HAuCl_4$	[128]
	Sargassum Wightii	Spherical	8-12	$HAuCl_4$	[114]
	Gelidiella Acerosa	spherical, hexagonal and cubical	5.81 - 117.59	$HAuCl_4$	[303]
Iron	*Sargassum muticum*	Cubic	18±4	$FeCl_3$	[134]
	Gracilaria edulis	Cubic	20-26	$FeCl_3$	[231]
	Chlorococcum sp.	Spherical	20 -50	$FeCl_3$	[230]
	Colpomenia sinuosa Pterocladia capillacea	Spheres	11.24–33.71 nm 16.85–22.47	$FeCl_3$	[138]
Copper	*Bifurcaria bifurcata*	Spherical	5-45	$CuSO_4$	

2.2.2. Algal Gold Nanoparticles

Use of gold for the treatment of various diseases was common in Egypt, India and China. Several organo-gold complexes with efficient antimicrobial, antitumor and antimalarial activities have developed in last few decades [84]. Extracellular biosynthesis of Au-NPs was reported in a comparatively short time span for the first time using *Sargassum wightii* with other biological procedures [114]. Currently there are several chemical and physical methods available for the preparation of metallic nanoparticles [115]. But alternative and environmentally safe process using biological systems for the production of NPs is a crucial need of time. There are number of reports available on the synthesis Au-NPs using microbes including algal members with many specified their activities for the control of pathogens [116-119] (Table 2).

Precipitation of gold nanoparticles using *Plectonema boryanum*, a filamentous cyanobacterium has been reported at membrane vesicles [120]. *Nostoc ellipsosporum* has been witnessed for the production of Au-NPs within the cells in laboratory condition at 20°C [121]. Biosynthesis of gold nanoparticles was investigated using three cyanobacterial strains *Leptolyngbya tenuis*, *Nostoc ellipsosporum* and *Coleofasciculus chthonoplastes*. In which *Coleofasciculus* showed maximum extracellular production of Au-NPs in medium unlike *Nostoc* and *Leptolyngbya* [122]. Using reversed-phase HPLC, isolated and purified protein (28 kDa) for reducing chloroauric acid [123]. *Chlorella vulgaris* was also observed as possible route for the bulk intracellular production of gold nanoparticles. NPs synthesized within the cells were recognized by Transmission electron microscopy and XRD. Higher than 97% gold nanoparticles was recovered from solution with a concentration of 1.4% Au using algal extract [94]. The biosynthesis of Au-NPs using *Stoechospermum marginatum* biomass was also reported as affective antimicrobial agent. Gold nanoparticles were achieved by the reduction of hydroxyl groups present in the diterpenoids of the algal extract [124]. Use of *Gracilaria corticata* to get Au-NPs was reported as an efficient, reliable process for the large-scale production of a new range of antimicrobial agents. Synthesis of Au-NPs exhibits rapid and non-toxic process having the remarkable potential in the biomedical field using algae extract [125]. Biosynthesis of Au-NPs using Turbinaria conoides extract was confirmed with yellow color change to deep pink. Crystalline structure of synthesized gold nanoparticles was studied using Transmission electron microscopy. Its crystalline nature is confirmed by X-ray diffraction. Because of the presence of natural constituents like polyphenolic substances and fucoidan, it eliminates the use of chemical substances [118, 126].

Au-NPs synthesized using *Corallina officinalis* extract, were almost 14.6 nm in diameter. Hydroxyl functional group present in the polyphenols and carbonyl group of algal proteins play fundamental role in the production of Au-NPs as revealed by FTIR analysis [127]. Both biologically synthesized Ag-NPs and Au-NPs A Antimicrobial potential of biologically synthesize using seaweeds extracts of *Padina tetrastromatica* and *Turbinaria ornate* was investigated through diverse analytical techniques [128].

2.2.3. Algal Iron Nanoparticles

As compared to silver and gold, available reports on the biosynthesis of iron nanoparticles using algae are very limited. Now it is well known that many species of seaweed hold potential medicinal applications against cancer, allergy, diabetes and number of other diseases [129-133]. Therefore, use of seaweeds can be good option for the large-scale production of biosynthetic nanoparticles as their metabolites contain important functional groups such as sulphate group, hydroxyl group, aldehyde, amino and carboxyl groups involve actively in the preparation of nanoparticles [134-135]. The improvement of methods using algal derived metabolites for nanoparticles is desire of time for its huge application in biomedical area as they are nontoxic, eco-friendly and reliable (Table 2).

Algal polysaccharides present within the extract having various functional groups provoke the conversion of Fe^{3+} present in the solution of ferric chloride its nanoparticles by the process of reduction. For the identification of these important groups present in the metabolites of algal extract FTIR spectroscopy was used which also confirmed that sulfate group actively participated in the production of Fe_3O_4-NPs by causing the reduction process [134]. In the marine alga *Porphyra vietnamensis* it was reported that sulphated polysaccharides had ability to synthesize nanoparticles by the conversion aldehyde groups to carboxylic acids through oxidation due to the important role of sulfate group [136]. The nanoparticle crystals were confirmed using TEM and X-ray powder diffraction technique showing good correlation between the particle sizes and crystalline structure of the iron nanoparticles [136-137].

Seaweed aqueous extracts Brown of *Colpomenia sinuosa* and *Pterocladia capillacea* were used to synthesise Fe_3O_4-NPs by reducing $FeCl_3$. Combinations of various available standard methods were used to characterize biosynthesized nanoparticles. Preliminary screening was done by observing the colour change during the reduction process, with *Colpomenia sinuosa* it was changed to brown and with *Pterocladia capillacea* it changed to buff indicating the synthesizing of Fe_3O_4-NPs [138]. Functional groups specifically sulphate group present in polysaccharides may help in the reduction of the ferric ions and caused the stabilization of the nanoparticles [139]. The FTIR spectrum of *Colpomenia sinuosa* and *Pterocladiella capillacea* depicted the major role of the sulfate group in the production of Fe_3O_4-NPs, as they disappeared as Fe_3O_4-NPs formed [60]. Biosynthesized Fe_3O_4-NPs utilizing queous extract *of Padina pavonica, Sargassum muticum* a and *Sargassum acinarium* was confirmed by EDX and UV spectra observing reading at 402 and 415 nm [134, 139]. The *Chaetomorpha antennina* green algal bioextract, as well as *Turbinaria turbinate* can give variations in the size, and morphology of Fe_3O_4-NPs at different pH, temperature, concentration of the precursor and algal extract [140].

2.2.4. Algal Copper Nanoparticles

Although researchers have focused mainly on silver, gold and iron nanoparticles now copper nanoparticles has drawn attention with properties of electrical, magnetic, thermic and antimicrobial nature Table 2. For the biosynthesis of copper nanoparticles (CO-NPs) aqueous extract of *Botryococcus braunii*, with copper acetate was used as precursor [101]. Color change from blue to brown resulted in the biosynthesis of copper nanoparticle, demonstrating the conversion of copper acetate into CO-NPs by the process of reduction [101, 141]. Scattering and absorption of biogenically synthesized metallic nanoparticles at a specific wavelength was involved in the reduction process of copper acetate [142].

Copper NPs synthesized by *Botryococcus braunii* exhibited the presence of the oxide shell around nanoparticles due to absorption peaks at 258 and 460nm [143-146]. These peaks (2-100 nm) have previously been reported for various metal nanoparticles [147-148].

Like Ag-NPs, stable biologically synthesized CO-NPs were also observed up to the period of 3 months in solution as depicted by the maintenance of brown color [146, 149]. There are major functional groups present in the *Botryococcus braunii* that were actively involved in the biogenic synthesis of copper nanoparticles by the process reduction as identified by using FTIR spectrum measurements. Scanning electron microscopy (SEM) for observing the shape and size of biosynthesized copper nanoparticles was used like other nanoparticles. Observed shape of copper-NPs was cubical and spherical with the diameter in the range of 10–70 nm [111, 143]. Biosynthesis of copper nanoparticles using algal extract was also confirmed by the analysis of X-ray diffraction (XRD) [143, 150]. Biosynthesis of Cu-NPs using *Sargassum polycystum* was obtained after the addition of its extract in the aqueous solution copper sulphate [151]. Mostly algae are considered as excellent source of diverse types of metabolites like polysaccharides, alkaloids, flavonoids, phenolic acids and carotenoids exhibited different activities responsible for the synthesis of Cu-NPs [151] (Table 2).

2.3. Fungal Nanoparticles

2.3.1. Fugal Silver Nanoparticles

Mycosynthesis of nanoparticles involves a bio-based reduction of different precursor salts with the help of various organic molecules produced by a variety of fungi. Many salts have been explored for fulfilling this requirement [152-153]. Silver presents an excellent opportunity for production of fungal based nanoparticles because of its various advantages. Different salts of silver when reacting with fungi produce a unique enzyme that converts metal ions to metal nanoparticles. Other metabolites are also produced which facilitate the overall reduction process [154-156]. Different fungal species have been employed for their potential to make stable nanoparticles [155] (Table 3).

Table 3. Production of different nanoparticles by fungal species

Nanoparticles	Fungi	Shape	Size (nm)	Source	References
Silver	*Aspergillu flavus*	spherical	8.92	$AgNO_3$	[167]
	A. fumigatus		–	$AgNO_3$	[168]
	A. terreus	spherical	1–20	$AgNO_3$	[166]
	Cladosporium cladosporioides		10–100	$AgNO_3$	[304]
	Coriolus versicolor	spherical	25–75 444–491	$AgNO_3$	[305]
	Fusarium oxysporum	spherical	20–50	$AgNO_3$	[164]
	Fusarium sp.		5–50nm	$AgNO_3$	[306]
	Macrophomina phaseolina	spherical	5–40	$AgNO_3$	[307]
	Penicillium brevicompactum		58.35	$AgNO_3$	[308]
	P. fellutanum	spherical	5–25	$AgNO_3$	[309]
	P. nalgiovense AJ12	spherical	25	$AgNO_3$	[310]
	Phaenerochaete chrysosporium	pyramidal	5–200	$AgNO_3$	[311]
	Phoma glomerata	-	60–80	$AgNO_3$	[312]
	Pleurotus ostreatus	spherical	< 40	$AgNO_3$	[163]
	P. sajor-caju	spherical	30.5	$AgNO_3$	[311]
	Trichoderma asperellum	Crystalline	13–18	$AgNO_3$	[154]
	Trichoderma harzianum	Cubic	51.10	$AgNO_3$	[248]
	T. viride	spherical	5–40	$AgNO_3$	[313]
Gold	*Aspergillus oryzae* var. viridis	*uniform in shape*	*10-60*	$HAuCl_4$	*[314]*
	Fusarium acuminatum	*spherical*	*8-28*	$HAuCl_4$	*[315]*

Nanoparticles	Fungi	Shape	Size (nm)	Source	References
	P. aurantiogriseum, Penicillium citrinum, Penicillium waksmanii	*Spherical*	153.3, 172, 160.1	AuCl	[316]
	Penicillium rugulosum.	*Spherical*	30–70	$HAuCl_4$	[317]
	Chrysosporium tropicum	-	2–15	$HAuCl_4$	[318]
	Candida albicans	*Spherical*	0–40	$HAuCl_4$	[319]
	Aspergillus clavatus	*Triangle*	20 - 35	$HAuCl_4$	[320]
Copper	*Stereum hirsutum*	Spherical	5 - 20	$CuCl_2$, $CuSO_4$, $Cu(NO_3)_2$	[195]
	Penicillium aurantiogriseum, Penicillium citrinum and *Penicillium waksmani*	Spherical	-	$CuSO_4$	[196]
	A. versicolor.	Round to polygonal shape	23–82	$CuSO_4$·	[321]
	Aspergillus niger	-	500	$CuSO_4$	[198]
Iron	*Aspergillus* sp	-	50-200	$FeSO_4$ $.5H_2O$	[322]
	Fusarium oxysporum and Verticillium sp	Spherical	20–50	$K_3[Fe(CN)_6]$ $K_4[Fe(CN)_6$	[323]
	P. chlamydosporium, A. fumigates, A. wentii, C. lunata and *C. globosum*	-	5–200	Fe_2O_3	[324]
	Alternaria alternate	cubic shape	9	$FeNO_3$	[188]
	(Trichoderma asperellum), STSP 19 (Phialemoniopsis ocularis) and *STSP 27 (Fusarium incarnatum)*	Spherical	25, 13.13, 30.56	$FeCl_3$ $FeCl_2$	[189]
	Aspergillus japonicus	Cubic	60–70	$K_3[Fe(CN)_6]$ $K_4[Fe(CN)_6]$	[191]

Trichoderma longibrachiatum was used for the production of silver nanoparticles. Fungal biomass was treated with silver salt ($AgNO_3$) for the bioreduction process. The brown color was produced after incubation at

28°C for 72-hours. Stable nanopartiples were produced which showed strong antifungal activity against phytopathogenic fungal species i.e., *Alternaria alternata, Pyricularia grisea, Fusarium verticillioides, Penicillium glabrum, P. brevicompactum, Aspergillus flavus, F. moniliform,* and *Helminthosporium oryzae* [157]. Spherical to ellipsoidal shaped silver nanoparticles were synthesized (79-107nm sizes). The size of the nanoparticle increased with the increase in the concentration of silver nitrate. These Ag-NPs delineated strong antimicrobial properties and cytotoxic properties [158].

Fusarium solani helped to reduce silver nitrate salt to form stable silver nanoparticles. Spherical shaped Ag-NPs with a size range of 5-35nm were obtained [159]. Another thermophilic fungal isolate *Humicola* sp facilitated the production of spherical silver nanoparticles. The fungal biomass was treated with $AgNO_3$ solution. After this treatment, the solution changed its color from yellow to brown due to surface plasmon vibrations. FTIR analysis confirmed the presence of proteins that ultimately formed stable nanoparticles. The size of these NPs was recorded between 5-25nm. These nanoparticles were also tested for cytotoxicity on cancer cell lines from human and mouse [160].

Hassan et al. [161] studied different factors which could optimize the formation of silver NPs using *Fusarium oxysporum.* Glucose and phosphate buffer were added for enhancing NPs efficiency. The maximum absorption λ_{max} was recorded at 430 nm after UV-visible spectroscopy. This study proved the involvement of microbial enzymes for production of nanoparticles and how changing precursor's concentration can affect the overall nanoparticles production process. Spherically shaped nanoparticles (25- 50 nm) were observed through SEM study.

Mushrooms are also studied for their nanoparticles production potential (Table 3). *Pleurotus cornucopiae* var. *citrinopileatus* were successfully employed in the production of silver nanoparticles. Mushrooms freeze-dried extract was combined with silver nitrate solution to obtain brownish solution due to the presence of polysaccharides. 20-30nm sized nanoparticles were observed that were spherical in the shape.

These Ag-NPs showed antifungal activity at higher NPs concentration (60μg/well) against various *Candida* sp [162].

Another oyster mushroom *Pleurotus ostreatus* was studied for its fungal mediated synthesis of Ag-NPs. The dried mushroom extract was treated with silver salt in dark for 6-40 hrs to obtain a dark brownish color solution. This change was the first indication of nanoparticles formation. 28nm sized NPs were obtained which were polydispersed [163]. Many other fungal species have been widely employed for the production of silver nanoparticles i.e., *Aspergillus sp, Fusarium sp, Metarhizium anisopliae, Trichoderma viride, etc.* [154, 156, 164-168]. Subramanian et al. [169] studied marine yeast (*Pichia capsulata*) for the synthesis of silver nanoparticles.

2.3.2. Fungal Gold Nanoparticles

Gold is also an important precursor for the fabrication of fungal nanoparticles. Gold NPs are shielded from oxidation [170]. The color and size of nanoparticles are interdependent. An array of colors (red, mauve, and yellow) corresponds to varying particle sizes [171]. Nanoparticles are stabilized by the addition of different compounds like ethyl alcohol and the bio-reduction process is carried out by enzymes secreted by fungal species, which prevent agglomeration of nanoparticles [172-173].

Different fungal species have been used in gold NPs production. Each specie helps in the fabrication of different shaped and sized NPs. Some are reported to possess strong antibacterial and antifungal properties [51, 165]. *Penicillium brevicompactum* was used for gold nanoparticles synthesis. These NPs attained their largest size at a higher concentration of gold salt. These gold NPs were also observed for their cytotoxic properties [174]. Different salts of gold like AuCl or $AuCl_3$ were used for the bioreduction process either in one or two steps [175].

Narayanan and Sakthivel [176] have studied fungus *Cylindrocladium floridanum for the production of gold nanoparticles. They obtained cubic shaped crystalline gold nanoparticles on the fungal surface. Verticillium* sp was used to synthesize nanoparticles by Mukherjee et al. [177]. Gold NPs were formed on the mycelial surface. AuNPs were entrapped on the fungal cell wall and cytoplasmic membrane. Anjana K [178] reported the formation of AuNps using a marine fungus *Aspergillus sydowii.* 10 nm diameters gold NPs were reported which were spherical and monodispersed in nature.

Zeinab et al. [179] studied fungus *Epicoccum nigrum* (isolation source: gold mine) for the production of gold nanoparticles. Fungal biomass was treated with chloroauric acid which led to the production of intracellular and extracellular nanoparticles. The size range of NPs was 5-50nm and they were spherical (Table 3).

In another study *Phanerochaete chrysosporium* was employed for the production of gold nanoparticles using gold chloride solution. Enzymes not only helped in stabilization but also reduction process. A deep purple color solution was obtained after the reaction at 37°C. Au ions helped to augment enzymatic activity. Laccase and ligninase enzymes were predominant in the solution. Nanoparticles were in 10-100nm in [180]. Different yeast species like *Hansenula anomala, Saccharomyces cerevisae,* and *Yarrowia lipolytica* were reported to synthesis gold nanoparticles under various conditions and produced diverse sized gold nanoparticles [181-183].

An endophytic fungal strain *Fusarium solani* ATLOY – 8 was studied for the production of gold nanoparticles. The anticancer potential of these nanoparticles was also studied. Different characterization methods were employed for observing the properties of NPs in detail. Nanoparticles size was between 40 and 45 nm diameter. These NPs delineated cytotoxicity in cervical cancer cells and human breast cancer cells [184].

In one study 29 fungal species were studied in detail for their gold nanoparticles production potential. Each specie produced diverse sized NPs. Bio AuNPs were in the size range of 1-80nm. Most of the nanoparticles were either spherical or hexagonal [185].

2.3.3. Fungal Iron Nanoparticles

Different Iron precursors have also been studied for the production of nanoparticles using various fungal species (Table 3). In one of the studies, *Fusarium oxysporum* and Verticillium sp, were used for iron nanoparticles synthesis utilizing mixtures of ferricyanide/ferrocyanide at room temperature. Quasi-spherical shaped NPs were obtained that were 20–50 nm in size. Cationic proteins from a fungal source facilitated the formation of stable iron oxide (Fe_3O_4) nanoparticles by breaking down anionic iron complexes. These proteins assisted in the capping and stabilizing process [186].

Kaul et al. [187] studied different fungal species like *P. chlamydosporium, A. fumigates, A. wentii, C. lunata and C. globosum* for the synthesis of iron nanoparticles. *C. globosum* helped in the formation of stable nanoaprticles after reaction with Fe_2O_3 solution. Each fungal sp synthesized different sized NPs i.e., *A. fumigatus* (42.4 nm). *C. lunata* (20.7nm), *C. globosum* (25 nm), and *A. wantii* (46 nm). Mohamed et al. [188] utilized *Alternaria alternate* fungus for iron nanoparticles production using iron nitrate as a precursor salt. The nanoparticles were cubic with size in the range of 5.4 to 12.1 nm. Synthesized NPs exhibited strong antibacterial activity against *E. coli, S. aureus,* and *P. aeroginosa* with a zone of inhibition of 13.2, 12.3, and 10.5 mm respectively.

Spherical shaped iron oxide nanoparticles were produced using three manglicolous fungal species (*Trichoderma asperellum*, *Phialemoniopsis ocularis,* and *Fusarium incarnatum*). Field emission scanning electron microscopy (FESEM) and TEM confirmed variable sizes of these iron nanoparticles i.e., *T. asperellum* (25 ± 3.94 nm), P. *ocularis* (13.13 ± 4.32 nm) and *F. incarnatum* (30.56 ± 8.68 nm) [189].

Proteins secreted by fungal species facilitate hydrolysis of main salt solution for stable nanoparticles production [154, 190]. *Aspergillus japonicus* isolate AJP01 was studied for production of iron oxide nanoparticles. The isolate successfully hydrolyzed iron cyanide complex. Cubic shaped IONPs were obtained with a size range of 60-70nm. The secreted proteins conferred stability to nanoparticles [191]. In another study, *Aspergillus niger* YESM 1 was employed for the synthesis of Magnetic Fe and Fe_3O_4 (magnetite) nanoparticles. Fungi helped in the breakdown of precursor iron salt to iron oxide. Ethanol was used in combination with nanoparticles at high temperatures. After analysis spherical nanoparticles were obtained with a size range of 18 to 50nm, these NPs had strong magnetic properties that could be utilized in biomedical applications [192].

Mycosynthesis of iron oxide nanoparticles from manglicolous fungi demonstrates a novel approach where mycosynthesized IO-NPs had shown effective results for the treatment of wastewater [193]. Fungus *Pleurotus* sp produced iron nanoparticles and it was speculated that it may be due to the presence of hydroxamates (iron transporting molecules)in fungi that bind to the complex iron molecules to take them inside of cells [194].

2.3.4. Fungal Copper Nanoparticles

Copper nanoparticles have been employed in various applications due to their properties. Many bacterial and plant species have made effective copper nanoparticles that are used mostly for their antibacterial activity. The white-rot fungus *Stereum hirsutum* was utilized to synthesize copper oxide/copper nanoparticles under varied pH conditions. Three different copper salts were used ($CuCl_2$, $CuSO_4$, and $Cu(NO_3)_2$). All salts were found effective for the synthesis of copper nanoparticles under alkaline conditions. After TEM analysis, spherically shaped (5 to 20nm) sized nanoparticles were confirmed. Extracellular proteins of fungi helped in nanoparticles formation as confirmed by FTIR analysis [195].

Penicillium species also contributed to the the formation of stable copper oxide nanoparticles (NPs). $CuSO_4$ salt was employed in different concentrations for the synthesis of nanoparticles. SEM analysis confirmed the presence of uniformly spherical nanoparticles. Each fungal species produced different sized copper NPs at different pH and salt concentrations. This study also proved the involvement of secreted proteins in the formation of stable nanoparticles, which helped to hydrolyze precursor metal salts to stable metal oxides [196].

Copper sulfide nanoparticles were synthesized using *Fusarium oxysporum* using an economical approach. Fungal biomass was treated with the copper sulfide solution, which led to the formation of black color from the initial violet color. Particle characterization was done following UV–vis spectroscopy, Fluorescence Spectroscopy, Fourier Transform Infrared Spectroscopy (FTIR), and Transmission Electron Microscopy

(TEM). Proteins helped in the synthesis of stable NPs by binding to nanoparticles. TEM analysis showed particles were round with an average size between 2-5nm and a diameter of 20nm [197].

Various fungal species have contributed to the formation of copper nanoparticles (Table 3). Common fungal specie, *Aspergillus niger* strain STA9 was successfully employed in the synthesis of copper nanoparticles. These Cu-NPs were further studied for their antibacterial and anticancer properties, 500 nm-sized Cu-NPs were synthesized. These nanoparticles showed cytotoxic properties towards carcinoma cell lines (Huh-7) with 3.09 μg/ml IC_{50} value. These Cu-NPs exhibited moderate antidiabetic properties. When these NPs were tested against various bacterial species they demonstrated antibacterial features against members of family Enterobacteriaceae besides other pathogens *ensuring the significance of* these NPs in the drug delivery [198].

3. Antibacterial Activities

3.1. Antibacterial Part of Plant Silver Nanoparticles

Antibiotic activities of nanoparticles synthesized from different plant species not only offer them as a suitable replacement candidature of antibiotics against resistance microbes of family Enterobacteriaceae but also as green formulations to adopt in pharmaceutical industries. This suggest that biogenic nanoparticles are more versatile biomedical products. Antibacterial activity of nanoparticles synthesized from *Ocimum tenuiflorum* and *Citrus sinensis* showed the average size of 28 nm, and 65 nm, respectively. Highest antimicrobial activity (well diffusion assay) was reported against different pathogenic strains (*S. aureus*) including members of family Enterobacteriaceae (*P. aeruginosa*, *E. coli* and *K. pneumoniae*) by *O. tenuiflorum* [199].

Silver nanoparticles were synthesized using *Carica papaya* showed the antibacterial activity of spherical shape nanoparticles (50–250 nm) against different gram positive and negative species and some from family *Enterobacteriaceae* [200].

Gnanadesigan et al. [201] reported the significance of leaves, roots and bark of mangrove plants having unique metabolites showing potent antibacterial activity of leaves extract against *E. coli (maximum zone size of inhibition)* and *Klebsiella* sp.

The efficacy of spherical to cuboidal shaped silver nanoparticles obtained from plant petal extracts of Cucurbita maxima, leaves of Moringa sp. and rhizome extract of *Acorus calamus* was tested. NPs showed efficacy against pathogenic bacterial isolates and cancer cell lines showing their promising results [202].

Among a wide variety of plants, succulents are also important for the synthesis of nanoparticles. For example, results of Geethalakshmi and Sarada, [40] study showed that aqueous root extract of *Trianthema decandra,* medicinal plant was able to synthesize gold and silver nanoparticles that were also effective against different pathogens from family Enterobacteriaceae e.g., *Enterococcus faecalis* (MTCC 2729), Escherichia coli (MTCC 443), *Pseudomonas aeruginosa* (MTCC 1035), *Proteus vulgaris* (MTCC 1771).

3.2. Antibacterial Effect of Plant Gold Nanoparticles

Preparation of AgNPs particles from leaf extract of *Solanum nigrum* (size 9.39 nm) showed antimicrobial activity effect against *E. coli.* This pathogen is least susceptible to antibiotics however, the polyphenols in plants bind with the microbial proteins and show antimicrobial effect [203]. In another study, *Euphorbia hirta* L leaves extract were used for the synthesis of gold nanoparticles and confirmed as a change in color from pale yellow to purple. These nanomaterials were synthesized in range from 6nm to 71nm, and characterization by TEM, XRD, EDAX, AFM, particle size analyzer, FTIR and Raman spectra also confirmed their presence. The efficacy of these gold nanoparticles was further confirmed when tested against different pathogenic strains like *E. coli*, *P. aeroginosa* and *K. pneumonia* [204].

To synthesize nanoparticles different organic solvents are used, however, using aqueous extract for synthesis of nanoparticles is more ecofriendly. Chandran et al. [205] reported the synthesis of AuNPs using aqueous extracts of *Cucurbita pepo* and *Malva crispa*. The nanoparticles showed antimicrobial activity against *E. coli* and *Listeria sp*. by disrupting cell membranes. *Ziziphus zizyphus* is a shrub with spines is a medicinal plant in Chinese culture, for their antifungal, antibacterial, antiulcer, and anti-inflammatory activities. Aljabali et al. [206] explained the synthesis of spherical shaped monodispersed gold nanoparticles of even 3 nm size (confirmed by TEM) showed no antibacterial or antifungal activity (of 5 mg/mL) indicating the biocompatible nature for the synthesized particles with least toxicity making them good source for using in drug delivery without disrupting commensal human microbiota.

3.3. Antibacterial Effect of Plant Iron Nanoparticles

Bacterial resistance to various antibiotics has been a matter of great concern however, green chemistry has offered a sparking substitute as antibacterial agents. Naseem and Farrukh, [207], prepared iron nanomaterial from two plant species as reducing compound i.e., *Lawsonia inermis* and *Gardenia jasminoides*. The response of these nanoparticles against different human pathogenic species were promising for example against *Escherichia coli*, *Salmonella enterica*, *Proteus mirabilis* and *Staphylococcus aureus*.

Iron nanoparticles have also been promising against different diseases Aziz, and Abd Urabe, [208] reported the efficacy of iron nanoparticles (20–99 nm) synthesized from *camellia sinensi* leaves and *Apium graveolens* extracts showed antibacterial activity against inhibited the growth of *S. aureus*, *P. aerugino*. Groiss et al. [209] reported the biosynthesized iron nanoparticles (spherical to crystalline shape Fe NPs) from *C. ramiflora* leaf extract effective to kill *E. coli* and *S. epidermidis*.

Aqueous extract from the leaves of *E. robusta* showed antimicrobial response against *Pseudomonas aeruginosa, Escherichia coli, Staphylococcus aureus* and *Bacillus subtilis*.

Sida cordifolia plant extract was also effective in synthesizing iron oxide nanoparticles which were found to be promising against different species of gram positive and gram negative bacteria [210]. Aqueous extract of *Musa ornata* flower extract used to synthesized iron nanoparticles (size -43.69 nm) were found to be effective against members of family Enterobacteriaceae and other human pathogens. Fe NPs were synthesized from *M. oleifera* leaf extract [211] with antimicrobial activity against pathogens *E. coli, P. aeruginosa*, *S. aureus*, *S. typhi* and *P. multocida* and speculated the reason of antibacterial activity that particles being positively charged area attached to negatively charged surfaces of bacteria and thus result in cell rupturing. Fe_3O_4 nanoparticle were synthesized from *C. guianensis* fruit and bactericidal action against different human pathogens was reported [212].

3.4. Antibacterial Effect of Plant Copper Nanoparticles

Copper nanoparticles with their diverse physiochemical properties have been synthesized using different methods (physical, chemical). However, biological synthesis has been in practice due to multiple environmental concerns. In spite of promising results against different pathogens Table 5 their practical implications is at initial stages requiring varying dose response experiments in environments. Being trace metal, copper have diverse applications with antimicrobial properties. From ancient times its antibacterial effect is known and different compounds have been in use as anti-microbial and antifungal agents against many pathogens including other members of family Enterobacteriaceae *S. aureus*, *S. enteric*, *C. jejuni*, *E. coli*, and *L. monocytogenes* [213]. The killing property of copper was reported due to extended contact killing time [214] and adhesion to cell wall associated proteins of gram positive either by denaturation of the intracellular proteins and interaction with macromolecules like DNA [215] and increases production of reactive oxygen species [216]. These NPs have been reported to be highly effective against antibiotic resistant *E. coli* strains. Kasana et al. [217] reported the efficacy and synthesis of copper nanoparticles (5-280 nm) using extracts from different plants like *Aloe vera, Bifurcaria bifurcate, Cassia alata, Syzygium aromaticum, Vitis vinifera, Centella asitica, Gloriosa superba and Citrus sp.* These biogenic copper nanoparticles have been reported to show promising antibacterial (*Enterococcus faecalis, Escherichia coli, Klebsiella pneumoniae, Shigella flexneri* besides other pathogens) and antifungal activity (*Fusarium* sp and and *Phytophthora infestans*). Similarly, extract from tea or coffee were potent enough to show antibacterial effect against different members of family Enterobacteriaceae and other human pathogens for example: *S. dysenteriae*, and *E. coli* [217].

Table 4. Antimicrobial activity of plant derived silver nanoparticles against pathogens of Enterobacteriaceae members

Nanoparticles	Biological entity	Test microorganisms	Method	References
Silver	*Carica papaya*	*Escherichia coli, Klebsiella pneumoniae*	Turbidimetry method	[200]
	Sesuvium portulacastrum	*Klebsiella pneumoniae*	Well diffusion	[269]
	Acalypha indica	*Escherichia coli*	Well diffusion	[270]
	Avicenna marina	*Escherichia coli, Klebsiella sp*	Disc diffusion	[201]
	Mentha piperita	*Ecoli*	Well diffusion	[278]
	Vitis vinifera	*Klebsiella planticolae*	Disc diffusion	[274]
	Ocimum tenuiflorum, Solanum tricobatum, Syzygium cumini, Centella asiatica Citrus sinensis	*Escherichia coli, Klebsiella pneumoniae*	Well diffusion	[199]

Table 4. (Continued)

Nanoparticles	Biological entity	Test microorganisms	Method	References
	Cucurbita maxima (petals), *Moringa oleifera* (leaves) and *Acorus calamus* (rhizome)	*Escherichia coli*	Well diffusion	[202]
	Cleome viscosa	*Escherichia coli, Klebsiella pneumoniae*	Well diffusion	[325]
	Trianthema decandra	*Enterococcus faecalis*, *Escherichia coli*, *Proteus vulgaris*, *Yersinia enterocolitica*	Disc diffusion	[40]
Gold	*Annona muricate*	*Entrococcus faecalis*	Well diffusion	[279]
	Solanum nigrum	*Escherichia coli*	Disc diffusion	[203]
	Ziziphus zizyphus Leaf extract	*Escherichia coli*	Radial diffusion	[206]
	Trianthema decandra	*Enterococcus faecalis*, *Escherichia coli*, *Proteus vulgaris*, *Yersinia enterocolitica*	Disc diffusion	[40]
	Mentha piperita	*Escherichia coli*	Well diffusion	[278]
	Euphorbia hirta	*Klebsiella pneumonia*	Broth dilution methods	[204]
	Salix alba	*Klebsiella pneumonia*	Agar well diffusion assay	[281]
Copper	*Citrus medicalinn*	*Propionibacterium acnes, Salmonella typhi, K. pneumoniae, Escherichia coli*	Disc diffusion	[298]
	Camellia sinensis	*Escherichia coli*	Disc diffusion	[326]
	Punica granatum	*Enterobacter aerogenes, Salmonella enterica*	Well diffusion	[327]

Nanoparticles	Biological entity	Test microorganisms	Method	References
	Tridax procumbens, Lantana camera, Azadirachta indica	*Escherichia coli*	Disc diffusion	[328]
	Citrus limon, Turmeric curcumin	*Escherichia coli, Enterobacter sp., Klebsiella pneumoniae Proteus spp. Salmonella Typhi*	Well diffusion	[329]
	Magnolia kobus	*Escherichia coli*	Broth dilution assay; CFUs	[330]
	Dodonaea viscosa	*Escherichia coli, K. pneumoniae*	Well diffusion	[61]
	Nerium oleander	*Escherichia coli, K. pneumoniae, Salmonella typhi*	Well diffusion	[331]
Iron	*Lawsonia inermis* and *Gardenia jasminoides*	*Escherichia coli, Salmonella enterica, Proteus mirabilis,*	Well diffusion	[207]
	Eucalyptus robusta	*Escherichia coli*	Disc diffusion	[332]
	Sida cordifolia	*Escherichia coli, Klebsiella pneumonia*	Well diffusion	[210]
	Musa ornata	*Escherichia coli, Salmonella enterica*	Well diffusion	[333]
	Erodium cicutarium	*Escherichia coli*	Dilution method	[334]
	L. camara	*Klebsiella*	Well diffusion	[335]
	Cynometra ramiflora	*Escherichia coli*	Dis diffusion	[209]
	Couroupita guianensis	*E. coli, S. typhi, K. penumoniae*	Disc diffusion	[212]
Gold	*Annona muricate*	*Entrococcus faecalis*	Well diffusion	[279]
	Solanum nigrum	*Escherichia coli*	Disc diffusion	[203]
	Ziziphus zizyphus Leaf extract	*Escherichia coli*	Radial diffusion	[206]
	Trianthema decandra	*Enterococcus faecalis, Escherichia coli, Proteus vulgaris, Yersinia enterocolitica*	Disc diffusion	[40]

Table 4. (Continued)

Nanoparticles	Biological entity	Test microorganisms	Method	References
	Mentha piperita	*Escherichia coli*	Well diffusion	[278]
	Euphorbia hirta	*Klebsiella pneumonia*	Broth dilution methods	[204]
	Salix alba	*Klebsiella pneumonia*	Agar well diffusion assay	[281]
Copper	*Citrus limon, Turmeric curcumin*	*Escherichia coli, Enterobacter sp., Klebsiella pneumoniae Proteus spp. Salmonella Typhi*	Well diffusion	[329]
	Magnolia kobus	*Escherichia coli*	Broth dilution assay; CFUs	[330]
	Dodonaea viscosa	*Escherichia coli, K. pneumoniae*	Well diffusion	[61]
	Nerium oleander	*Escherichia coli, K. pneumoniae, Salmonella typhi*	Well diffusion	[331]
Iron	*Lawsonia inermis* and *Gardenia jasminoides*	*Escherichia coli, Salmonella enterica, Proteus mirabilis*	Well diffusion	[207]
	Eucalyptus robusta	*Escherichia coli*	Disc diffusion	[332]
	Sida cordifolia	*Escherichia coli, Klebsiella pneumonia*	Well diffusion	[210]
	Musa ornata	*Escherichia coli, Salmonella enterica*	Well diffusion	[333]
	Erodium cicutarium	*Escherichia coli*	Dilution method	[334]
	L. camara	*Klebsiella*	Well diffusion	[335]
	Cynometra ramiflora	*Escherichia coli*	Dis diffusion	[209]
	Couroupita guianensis	*E. coli, S. typhi, K. penumoniae*	Disc diffusion	[212]

Table 5. Antimicrobial activity of algal silver nanoparticles against pathogens of Enterobacteriaceae members

Nanoparticles	Biological entity	Test microorganisms	Method	References
Silver	*Chaetoceros calcitrans, Cuscuta salina, Isochrysis galbana*	*Escherichia coli, Klebsiella* sp, *Proteus* sp, *Pseudomonas* sp	-	[223]
	Urospora sp.	*Staphylococcus aureus, Escherichia coli* and *Pseudomonas aeruginosa.*	Well diffusion	[224]
	Pithophora oedogonia	*Escherichia coli, Pseudomonas aeruginosa, Vibrio cholera, Shigella flexneri, Bacillus subtilis, Staphylococcus aureus, Micrococcus luteus*	Well diffusion	[111]
	Padina tetrastomatica	*Staphylococcus aureus, Bacillus subtilis, Pseudomonas aeruginosa, Salmonella typhi, Escherichia coli*	disc diffusion	[97]
	Sargassum ilicifolium	*Klebsiella pneumoniae, Pseudomonas aeruginosa, Staphylococcus aureus, Escherichia coli, Enterococcus faecalis, Proteus mirabilis*	Disc diffusion	[225]
	Gracilaria birdiae	*Staphylococcus aureus, Escherichia* coli	96-Well microdilution	[302]
	Sargassum tenerrimum	*Bacillus cereus, Escherichia coli, Klebsiella pneumoniae, Proteus mirabilis, Pseudomonas aeruginosa, Salmonella typhii,*	Disc diffusion	[124]
	Caulerpa racemosa	*Staphylococcus aureus, Proteus mirabilis*	Well diffusion	[336]

Table 5. (Continued)

Nanoparticles	Biological entity	Test microorganisms	Method	References
Gold	*Turbinaria conoides*	*Bacillus subtilis, Klebsiella pneumonia* and *Streptococcus* sp	Well diffusion	[126]
	Galaxaura elongata	*E. coli, K. pneumoniae* and MRSA, respectively, followed by *S. aureus* and *P. aeruginosa*	Well diffusion	[337]
	Gracilaria corticata	*Enterobacter aerogenes, Escherichia coli*	Well diffusion	[125]
	Gelidiella acerosa	*Escherichia coli, Serratia marcescens, Klebsiella pneumonia, Bacillus subtilis*	Well diffusion	[119]
Iron	*Colpomenia sinuosa, Pterocladia capillacea*	*Escherichia coli, Pseudomonas aeruginosa, Salmonella typhi, Vibrio cholera, Bacillus subtilis, Staphylococcus aureus*	Disc diffusion	[138]
	Gracilaria edulis	*Bacillus cereus, Escherichia coli, Klebsiella pneumoniae, Pseudomonas aeruginosa, Salmonella typhimurium*	Well diffusion	[231]
Copper	*Bifurcaria bifurcata*	*Enterobacter aerogenes, Staphylococcus aureus*	disc diffusion	[143]
	Botryococcus braunii	*Pseudomonas aeruginosa, Escherichia coli, Klebsiella pneumoniae, Staphylococcus aureus*	Agar well plate	[101]
	Sargassum polycystum	*Escherichia coli, Pseudomonas aeruginosa, Salmonella typhimurium, Vibirio cholerae*	Well diffusion	[151]

3.5. Antibacterial Effect of Algal Silver Nanoparticles

Silver is non-toxic and has been used over the past centuries for the control of pathogenic microbes as well as an attractive antimicrobial agent (Table 5) [218]. Because of its ability to inhibit the wide range of antibiotic-resistant bacteria even at minute concentration, silver ions, are extensively used for various medical purposes [219]. Silver nanoparticles synthesized using algae are safe for the environment as they do not involve any toxic chemicals in its synthesis [88]. Silver nanoparticles served as an efficient protective barrier for most of the microorganism including pathogens [87, 220]. Toxic effects of synthesized silver nanoparticles on human pathogen creates avenues for a noval antimicrobial agents [221]. Silver nanoparticle synthesized using *Padina sp.* and *Hypnea* sp. extracts extract exhibited efficient antibacterial activity against gram-positive bacteria and gram-negative bacteria [222]. Silver nanoparticles synthesized using *Sargassum tenerrimum* showed remarkable activity against nine pathogens. Silver compounds as positively charged nanoparticles resulted in cell death while interacting with negatively charged bacterial membrane [110]. Ag-NPs synthesis using *Chaetoceros calcitrans, Cuscuta salina, Isochrysis galbana,* Urospora sp. *Pithophora oedogonia* showed antimicrobial activity againt large number of gram negative and gram positive bacteria [111, 223, 224]. Disc impregnated with the Ag-NPs synthesized by *Sargassum ilicifolium*, *Sargassum tenerrimum* and *Padina tetrastomatica*, were evaluated for their antimicrobial activities showed remarkable control of pathogens [87, 110, 225]. Detail and exact mechanism involve in its antibacterial activities is not clear yet but most of the researches have suggested that Ag-NPs may attack the cell membrane and impair its permeability [226]. Some studies predicted that nanoparticles damaged DNA after penetrating into the cell. Nanoparticles also provide larger surface area to the volume ratio because of its small size [227].

3.6. Antibacterial Effect of Algal Gold Nanoparticles

Algal derived compounds (Table 5) and their role in synthesising gold nanoparticles using *Turbinaria conoides* was confirmed visually by observing change in colour from brown to pinkish red [126]. Biologically synthesized Au-NPs act as potent antibacterial agent against *Klebsiella pneumoniae* that is an important human pathogen. Antibacterial studies depicted by biogenic synthesized Au-NPs using *Galaxaura elongata* was significantly higher than powder Au-NPs or free ethanolic extract [88]. It may be due the presence of algal secondary metabolites. Gold ions inactivate the bacterial cells after reacting with SH groups of proteins [228]. Proteomic analysis showed that Au-NPs caused alteration in gene expression of Escherichia coli even at short period of exposure [229]. Nanoparticles effect the bacterial cell permeability by disturbing the structure of cell membrane [88, 125]. Like Ag-Nps, Au-NPs also have more harmful impact on gram negative organisms as they have thin cell wall with few layers of peptidoglycans allow less resistance against the penetration of nanoparticles [230].

3.7. Antibacterial Effect of Algal Iron Nanoparticles

Antibacterial potential of Fe_3O_4 nanoparticles Table 5 synthesized using *Gracilaria edulis* was observed against the growth of *Pseudomonas aeruginosa* and *Klebsiella pneumoniae* [231]. The detrimental effect of iron oxide nanoparticles on the growth of microbes was attributed to their smaller size as they can easily penetrate the cell membrane and disturbed its permeability [58]. The biogenically synthesized Fe_3O_4-NPs using *Colpomenia sinuosa* and *Pterocladia capillacea* demonstrated wide range inhibition spectrum against both gram positive and negative bacteria as compared with the ampicillin and its pure aqueous extracts. But the antibacterial effect of Fe_3O_4-NPs against gram negative than gram positive bacteria was remarkable higher in ratio [138]. It may be due the difference in position of their cell wall as gram positive bacteria is composed of more peptidoglycan layer provide less resistance against the penetration of Fe_3O_4-NPs into the cell whereas the cell wall of gram negative bacteria has comparatively few peptidoglycan layer make them more vulnerable to Fe_3O_4-NPs [228]. Antimicrobial potential of Fe_3O_4-NPs also depends on the reaction take place between the protein's thiol groups present existing on the surface of the bacterial cell ions released from the nanomaterials [232].

Small sized particles like cubic and spherical nanoparticles offer high surface to volume ratio gave better opportunity to interact with the microbes as compared to large sized particles [233]. Antimicrobial property of the algal (*Dictyota dicotoma*) synthesized Fe_3O_4-NPs against different pathogens [234]. Antimicrobial activity of iron nanoparticles is correlated to its small sizes [235]. Synthesis of free radicals also add in the antimicrobial potency of Fe_3O_4-NPs during reaction [236] besides the reported electromagnetic force of attraction between charge differences of nanoparticles and microbes [237].

3.8. Antibacterial Effect of Algal Copper Nanoparticles

Copper nanoparticles synthesized by using *Botryococcus braunii* showed inhibitory effects against all tested organisms i.e., S. *aureus*, *K. pneumoniae*, *E. coli* and *P. aeruginosa* [88]. Inhibitory effect of CuNPs Table 5 may be due subsequently release of copper ions which may denature the DNA molecules and disrupt biochemical processes within the bacterial cells [238-239]. The antibacterial activity observed using *Bifurcaria bifurcate* extract and its copper nanoparticles showed that discs prepared with only with algal extract did not showed any inhibitory effect but CuNPs synthesized from its extract exhibited remarkable antibacterial activity against pathogens [240]. CuNPs also exhibited effective antimicrobial activity due to the provision of large surface area for the enhanced interaction with targeted microorganisms. The behavior observed by the biologically synthesizof Cu-NPs showed that antimicrobial effect was more pronounced on gram positive bacteria than gram-negative and in line with another previous findings [241]. It was due to basic structural differences in the cell wall of *Enterobacter aerogenes* and *Staphylococcus aureus* [242]. Cu-NPs synthesized by using brown seaweed *Sargassum polycystum* using exhibited efficient antimicrobial activity against Pseudomonas aeruginosa and *Shigella dysenteriae* [151].

3.9. Antibacterial Effect of Fungal Silver Nanoparticles

Nanoparticles have been explored for various applications in different fields from electronics to biomedical [243]. One of the most efficient property is their antibacterial nature [244-245]. Nanoparticles exhibit this property due to their nanoscale size which ensures more surface area for bacterial binding which ultimately decreases the oxygen supply to microbes. The size determines their extent of antimicrobial properties as smaller sized NPs are deemed more potent against bacteria. Fungal nanoparticles are easier to handle and can be easily grown in lab conditions. This not only provides a green route of nanoparticles fabrication but also reduces cost of NPs synthesis [156] (Table 6).

Silver nanoparticles were fabricated using *Talaromyces purpureogenus* mycelial extract to check their antibacterial potential. These were characterized using traditional analysis methods. These round-shaped nanoparticles significantly decreased the growth of various bacterial pathogens like *E. coli* and *P. aeruginosa* but Ag-NPs had the less pronounced antibacterial effect in *S. enterica* [246].

Aspergillus terreus was exploited for the synthesis of silver nanoparticles. These Ag-NPs were tested against 12 bacterial strains. The highest antibacterial was recorded in *S. typhi* followed by *E. coli* and then *S. marcescens* (15.33 ± 0.58 mm). *K. pneumoniae* (MDR) had the lowest zone of inhibition. MIC values vary among tested strains i.e., *S. typhi, E. coli, S. aureus, S. flexneri,* and *S. marcescens* (11.43 µg/ml), *P. aeruginosa* (32.2 µg/ml), for *P. mirabilis and E. faecalis* (102 µg/ml) and *K. pneumoniae* (308 µg/ml) [247].

Ahluwalia et al. [248] fabricated Ag-NPs utilizing *Trichoderma harzianum* to study their antibacterial properties. These NPs had more activity against *K. pneumoniae* with a large zone of inhibitions as compared to gram positive bacteria. Ag-NPs helped in the distortion of the plasma membrane which enhanced the antibacterial effect. Silver nanoparticles were made from *A. niger* by Gade et al. [159] also demonstrated antibacterial activity.

Table 6. Antibacterial properties of various mycosynthesized nanoparticles

Nanoparticles	Fungal Source	Test microorganisms	Method	References
Silver	*Macrophomina phaseolina*	*Escherichia coli (DH5α),*	Disc diffusion method	[307]
	Penicillium sp.	*E. coli, S. aureus*	Agar well diffusion assay method	[254]
	Pleurotus ostreatus	*Escherichia coli, Klebsiella pneumoniae, Pseudomonas aeruginosa, Staphylococcus aureus* and *Vibrio cholera*	Disc diffusion method	[340]
	Trichoderma crassum	*E.coli (DH5α), Agrobacterium tumifaciens (LBA4404), Magnaporthe oryzae*	Disc diffusion method	[341]
	Candida glabrata	*S. aureus, E. coli, S. typhimurium, S. flexneri, K. pneomonaie, P. aeruginosa,* and *Candida* spp., e.g., *C. albicans, C. dubliniensis, C. parapsilosis, C. tropicalis, C. krusei,* and *C. glabrata*	Well diffusion	[342]
	A. niger	*Escherichia coli*	Agar diffusion assay	[338]
	Fusarium acuminatum	*Escherichia coli, Salmonella typhi (MTCC-733),*	Well diffusion	[339]
	Fusarium oxysporum	*Escherichia coli, Enterobacter cloacae, Klebsiella pneumoniae, Proteus mirabilis, Pseudomonas aeruginosa, Staphylococcus aureus,* and *Staphylococcus epidermidis*	Growth inhibition assays	[343]
	Aspergillus terreus	*Aspergillusflavus, Aspergillus fumigates, Aspergillusniger, Candida albicans* and *Staphylococcus aureus*	Disc diffusion method	[344]
	Ganoderma sessiliforme	*Escherichia coli, Bacillus subtilis, Streptococcus faecalis, Listeria innocua* and *Micrococcus luteus*	Disc diffusion assay	[345]

Table 6. (Continued)

Nanoparticles	Fungal Source	Test microorganisms	Method	References
	Penicillium aculeatum Su1	*Escherichia coli (E. coli, ATCC-8739), Pseudomonas aeruginosa (P. aeruginosa, ATCC-15442*	Kirby-Bauer disc diffusion method	[346]
	Fusarium oxysporum	*Escherichia coli* and *Staphylococcus aureus*	Disc diffusion assay	[255]
Copper	*Aspergillus niger*	*Escherichia coli, Staphylococcus aureus, Klebsiella pneumoniae, Micrococcus luteus* and *Bacillus subtilis*	Disc diffusion assay	[198]
Gold	*Cladosporium cladosporioides*	*E. coli MTCC 118, Staphylococcus aureus MTCC 7443, Bacillussubtilis MTCC 441, Pseudomonas aeruginosa MTCC 424,* and a fungal pathogen, *Aspergillus nigerMTCC 28*	Disc diffusion assay	[259]
	Pleurotus ostreatus	*E. coli, Staphylococcus aureus*	Disc diffusion assay	[257]
	Trichoderma harzianum	*E. coli MTCC 1302*	Disc diffusion assay	[260]
	Rhodotorula sp. *ATL72*	*Streptococcus sp., Bacillus* sp., *Staph* sp., *Shigella* sp., *Escherichia coli, Pseudomonas aeruginosa, Klebsiella* sp. and *Candida* sp.	MIC assay	[347]
Iron	*Alternaria alternate*	*Bacillus subtilis* than *Escherichia coli, Staphylococcus aureus* and *Pseudomonas aeruginosa*	Disc diffusion assay	[188]
	Fusarium Oxysporum	*Staphylococcus aureus, Klebsiella pneumoniae, Proteus vulgaris, Pseudomonas aeruginosa* and *Escherichia coli.*	Disc diffusion assay	[263]

Silver nanoparticles using *Candida albicans* depicted remarkable antibacterial activity against *S. aureus, B. cereus, E. coli, V. cholerae*, and *P. vulgaris*. [249]. The antibacterial effect was augmented with the combination of silver NPs and antibiotics. Similar studies were conducted using fungal species named *A. flavus, A. niger* and *T. harzianum* that backed the fact that silver nanoparticles have efficient antimicrobial properties against *B. cereus, B. subtilis, E. aerogenes, E. coli*, and *S. aureus* [250, 251]. *Trametes ljubarsky* and *Ganoderma enigmaticum* mediated silver nanoparticles effectively controlled the growth of pathogens like *B. megaterium, B. subtilis, S. aureus, B. cereus, M. luteus, E. aerogens, K. pneumoniae, E. coli, S. typhimurium, P. vulgaris, P. aeruginosa*, and *S. paratyphi* [252].

Silver nanoparticles have also been used to control phytopathogenic bacterial species in cotton. *Alternaria alternate* was used for silver nanoparticles syntheses that were capped with butyl acrylate. These Ag-NPs inhibited 99.9% growth of *E. coli* and *S. aureus* [253]. Mycosynthesis of silver NPs using *Penicillium* sp. showed promising results for controlling multidrug-resistant *E. coli* and *S. aureus* [254]. Moreover, silver NPs can also be used for its anti-cancerous properties. *Fusarium oxysporum* helped in the bioreduction of silver salts to produce potent NPs that helped in the control of *E. coli* and *S. aureus*. They also had antitumor properties [255].

3.10. Antibacterial Effect of Fungal Gold Nanoparticles

Gold nanoparticles (AuNPs) are an excellent choice for studying antibacterial properties. *Lignosus rhinocerotis* sclerotial extract and chitosan were used in the synthesis of gold nanoparticles using chloroauric acid ($HAuCl_4$). These AuNPs depicted antibacterial activity against tested bacterial species i.e., *P. aeruginosa, E. coli, S. aureus,* and *Bacillus* sp. [256].

Many other fungal species-based nanoparticles have reported antibacterial activity against both gram-positive and gram-negative bacteria. *Pleurotus ostreatus* based gold nanoparticles were found effective against *E. coli* and *S. aureus* [257]. Similar results were reported for

A. terreus IF0 based gold nanoparticles that had bactericidal effects against the pathogenic gram-negative bacteria, *E. coli* [258].

The synthesis of nanoparticle was dependent on the activity of NADPH-reductase and the presence of phenolic compounds. These Au-Nps depicted showed antibacterial activity against tested species *E. coli, S. aureus, B. subtilis* and *P. aeruginosa* [259]. Biologically synthesized Au-NPs using *T. harzianum* biomass showed effective antibacterial activity [260] (Table 6).

Sixty-five fungal strains were tested for their nanoparticles making ability and antimicrobial features. *Aspergillus terreus* was successful in making stable NPs, which also exhibited antibacterial features against *Staphylococcus aureus* and *Bacillus subtilis* [261]. Yeast based nanoparticles also have antibacterial properties. Au-NPs and Ag-NPs prepared using *Candida guilliermondii* [182], demonstrated a strong antibacterial effect against *S. aureus*.

Aspergillus niger was used in combination with a gold precursor for the formation of gold nanoparticles. These nanoparticles had lower activity against *E. coli*, *K. pneumonia*, *P. aeruginosa*, *S. aureus*, and *V. cholera* as compared to antibiotics, But in combination with antibiotics, their antibacterial effects were amplified [262].

3.11. Antibacterial Effect of Fungal Iron Nanoparticles

Iron nanoparticles have diverse activity against many bacterial species [207]. Fungal species *Alternaria alternata* was used for iron NPs synthesis. Gram-positive and gram-negative bacteria (*B. subtilis* than *E. coli, S. aureus,* and *P. aeruginosa)* were controlled using these Fe-NPs Table 6. These NPs caused a severe oxidative damage to bacterial cell which led to the death of cells [188]. *F.oxysporum* was employed for iron nanoparticle formation. These nanoparticles effectively inhibit the growth of *Bacillus, E. coli,* and *Staphylococcus* sp. [263].

3.12. Antibacterial Effect of Fungal Copper Nanoparticles

Copper nanoparticles (Cu-NPs) are routinely used for their antibacterial properties [215] Table 6. Synthesis of Cu-NPs using *A. niger* was evaluated for their antimicrobial and antitumor properties. These Cu-NPs delineated antibacterial activity against various bacterial species like *B. subtilis*. *E. coli, S. aureus, K. pneumoniae and M. luteus* [198].

Conclusion

In the current review, insights of the synthesis of nanoparticles from biological sources (plants, bacteria and fungi) have been provided. The review summarizes that in spite of multiple available options for the synthesis of nanoparticles, biologically derived synthesis is more reliable and promising than other available methods. This potential is described in terms of its cost efficacy, its eco-friendly and sustainable features and multiplication at industrial levels. The syntheses of nanoparticles are most profiting in the field of biological sciences where the most resistant bacteria from family Enterobacteriaceae are targeted.

However, to sum up it is important to describe here that this multifaceted research field of biosynthesized nanoparticles become more crucial because the development of size and shape of nanoparticles are also important besides the green synthesis. As a suitable carrier or as drug delivery systems, their non toxicity and extending the current antibacterial effect range against members of family Enterobacteriaceae will make the nanoparticles further showing ideal candidature in the field of biomedicine.

Future Perspectives

To sum up the current review, the family Enterobacteriaceae has been of great significance from medical point of view. The development of antibiotic resistance in members of family Enterobacteriaceae led us to rethink on our future endeavors to switch towards the green synthesis of nanoparticles instead of our traditional approaches to cope with the members. In this regard, the utilization of unexplored but sufficient resources of plants, algae and fungi should be focused. Since the diverse applications of biogenic nanoparticles is sparking the thoughts of industrialist, however, being scientists cost efficacy and zero toxicity should be the basic criteria of our research input.

REFERENCES

[1] Nordmann, P. (2014). Carbapenemase-producing Enterobacteriaceae: overview of a major public health challenge. *Medecine et maladies infectieuses*, 44 (2): 51-56].

[2] Bar-Oz, B., Preminger, A., Peleg, O., Block, C. and Arad, I. (2001). *Enterobacter sakazakii* infection in the newborn. *Acta Paediatrica*, 90 (3): 356-358].

[3] Safdar, N. and Maki, D. G. (2006). Use of vancomycin-containing lock or flush solutions for prevention of bloodstream infection associated with central venous access devices: a meta-analysis of prospective, randomized trials..*Clinical infectious diseases*, 43 (4): 474-484].

[4] Mayhall, C. G., Lamb, V. A., Gayle Jr, W. E. and Haynes Jr, B. W. (1979). *Enterobacter cloacae septicemia* in a bum center: epidemiology and control of an outbreak. *Journal of Infectious Diseases*, 139 (2): 166-171].

[5] Woo, A. H., Goetz, A. and Victor, L. Y. (1992). Transmission of Legionella by respiratory equipment and aerosol generating devices. *Chest*, 102 (5): 1586-1590].

[6] Baden, L. R. and Eisenstein, B. I. (2000). Impact of antibiotic resistance on the treatment of Gram-negative sepsis. *Current infectious disease reports*, 2 (5): 409-416].

[7] Breijyeh, Z., Jubeh, B. and Karaman, R. (2020). Resistance of Gram-negative bacteria to current antibacterial agents and approaches to resolve it. *Molecules*, 25 (6): 1340].

[8] Reyes, V. C., Opot, S. O. and Mahendra, S. (2015). Planktonic and biofilm-grown nitrogen-cycling bacteria exhibit different susceptibilities to copper nanoparticles. *Environmental toxicology and chemistry*, 34 (4): 887-897].

[9] Wang, L., Hu, C. and Shao, L. (2017). The antimicrobial activity of nanoparticles: present situation and prospects for the future. *International journal of nanomedicine*, 12: 1227].

[10] Ramalingam, B., Parandhaman, T. and Das, S. K. (2016). Antibacterial effects of biosynthesized silver nanoparticles on surface ultrastructure and nanomechanical properties of gram-negative bacteria viz. *Escherichia coli* and *Pseudomonas aeruginosa*. *ACS applied materials & interfaces*, 8 (7): 4963-4976].

[11] Samuel, U. and Guggenbichler, J. (2004). Prevention of catheter-related infections: the potential of a new nano-silver impregnated catheter. *International Journal of Antimicrobial Agents*, 23: 75-78].

[12] Galiano, K., Pleifer, C., Engelhardt, K., Brössner, G., Lackner, P., Huck, C., Lass-Flörl, C. and Obwegeser, A. (2008). Silver segregation and bacterial growth of intraventricular catheters impregnated with silver nanoparticles in cerebrospinal fluid drainages. *Neurological research*, 30 (3): 285-287].

[13] Sanzari, I., Leone, A. and Ambrosone, A. (2019). Nanotechnology in plant science: to make a long story short. *Frontiers in Bioengineering and Biotechnology*, 7: 120].

[14] Baskar, V., Meeran, I., Subramani, A. and Sruthi, A. J., Shabeer, TK 2018. Historic review on modern herbal nano gel formulation and delivery methods. *International Journal of Pharmacy and Pharmaceutical Sciences*, 10 (10).

[15] Rastogi, A., Tripathi, D., Yadav, S., Chauhan, D., Živčák, M., Ghorbanpour, M., El-Sheery, N. and Brestic, M. (2019). Application of silicon nanoparticles in agriculture. 3 *Biotech* 9 (3), 90. Nafees et al.

[16] Viswanath, B. and Kim, S. 2016. "Influence of nanotoxicity on human health and environment: the alternative strategies." In *Reviews of environmental contamination and toxicology*, 61-104. Springer.

[17] Kaviya, S., Santhanalakshmi, J., Viswanathan, B., Muthumary, J. and Srinivasan, K. (2011). Biosynthesis of silver nanoparticles using *Citrus sinensis* peel extract and its antibacterial activity. *Spectrochimica Acta Part A: Molecular and Biomolecular Spectroscopy*, 79 (3): 594-598].

[18] Makarov, V., Love, A., Sinitsyna, O., Makarova, S., Yaminsky, I., Taliansky, M. and Kalinina, N. (2014). "Green" nanotechnologies: synthesis of metal nanoparticles using plants. *Acta Naturae (англоязычная версия)*, 6 (1 (20)).

[19] Velayutham, K., Rahuman, A. A., Rajakumar, G., Roopan, S. M., Elango, G., Kamaraj, C., Marimuthu, S., Santhoshkumar, T., Iyappan, M. and Siva, C. (2013). Larvicidal activity of green synthesized silver nanoparticles using bark aqueous extract of *Ficus racemosa* against *Culex quinquefasciatus* and *Culex gelidus*. *Asian Pacific Journal of Tropical Medicine*, 6 (2): 95-101].

[20] Lara, H. H., Ayala-Núñez, N. V., Turrent, L. d. C. I. and Padilla, C. R. (2010). Bactericidal effect of silver nanoparticles against multidrug-resistant bacteria. *World Journal of Microbiology and Biotechnology*, 26 (4): 615-621].

[21] Fayaz, A. M., Girilal, M., Mahdy, S. A., Somsundar, S., Venkatesan, R. and Kalaichelvan, P. (2011). Vancomycin bound biogenic gold nanoparticles: a different perspective for development of anti VRSA agents. *Process biochemistry*, 46 (3): 636-641].

[22] Furno, F., Morley, K. S., Wong, B., Sharp, B. L., Arnold, P. L., Howdle, S. M., Bayston, R., Brown, P. D., Winship, P. D. and Reid, H. J. (2004). Silver nanoparticles and polymeric medical devices: a new approach to prevention of infection? *Journal of Antimicrobial Chemotherapy*, 54 (6): 1019-1024]

[23] Kumar, P. V., Pammi, S., Kollu, P., Satyanarayana, K. and Shameem, U. (2014). Green synthesis and characterization of silver

nanoparticles using *Boerhaavia diffusa* plant extract and their anti bacterial activity. *Industrial Crops and Products*, 52: 562-566].

[24] Ahmed, S., Ahmad, M., Swami, B. L. and Ikram, S. (2016). A review on plants extract mediated synthesis of silver nanoparticles for antimicrobial applications: a green expertise. *Journal of advanced research*, 7 (1): 17-28].

[25] Panáček, A., Smékalová, M., Večeřová, R., Bogdanová, K., Röderová, M., Kolář, M., Kilianová, M., Hradilová, Š., Froning, J. P. and Havrdová, M. (2016). Silver nanoparticles strongly enhance and restore bactericidal activity of inactive antibiotics against multiresistant Enterobacteriaceae. *Colloids and Surfaces B: Biointerfaces*, 142: 392-399].

[26] Mourdikoudis, S., Pallares, R. M. and Thanh, N. T. (2018). Characterization techniques for nanoparticles: comparison and complementarity upon studying nanoparticle properties. *Nanoscale*, 10 (27): 12871-12934].

[27] Khan, T., Ullah, N., Khan, M. A. and Nadhman, A. (2019). Plant-based gold nanoparticles; a comprehensive review of the decade-long research on synthesis, mechanistic aspects and diverse applications. *Advances in colloid and interface science*, 272: 102017].

[28] Verissimo, T. V., Santos, N. T., Silva, J. R., Azevedo, R. B., Gomes, A. J. and Lunardi, C. N. (2016). In vitro cytotoxicity and phototoxicity of surface-modified gold nanoparticles associated with neutral red as a potential drug delivery system in phototherapy. *Materials Science and Engineering: C*, 65: 199-204].

[29] Wiley, B., Sun, Y. and Xia, Y. (2007). Synthesis of silver nanostructures with controlled shapes and properties. *Accounts of Chemical Research*, 40 (10): 1067-1076].

[30] Sánchez-Navarro, M. d. C., Ruiz-Torres, C. A., Niño-Martínez, N., Sánchez-Sánchez, R., Martínez-Castañón, G. A., DeAlba-Montero, I. and Ruiz, F. (2018). Cytotoxic and bactericidal effect of silver nanoparticles obtained by green synthesis method using *annona*

muricata aqueous extract and functionalized with 5-fluorouracil. *Bioinorganic chemistry and applications*, 2018.

[31] Burygin, G., Khlebtsov, B., Shantrokha, A., Dykman, L., Bogatyrev, V. and Khlebtsov, N. (2009). On the enhanced antibacterial activity of antibiotics mixed with gold nanoparticles. *Nanoscale research letters*, 4 (8): 794-801].

[32] Payne, J. N., Waghwani, H. K., Connor, M. G., Hamilton, W., Tockstein, S., Moolani, H., Chavda, F., Badwaik, V., Lawrenz, M. B. and Dakshinamurthy, R. (2016). Novel synthesis of kanamycin conjugated gold nanoparticles with potent antibacterial activity. *Frontiers in microbiology*, 7: 607].

[33] Rattanata, N., Klaynongsruang, S., Leelayuwat, C., Limpaiboon, T., Lulitanond, A., Boonsiri, P., Chio-Srichan, S., Soontaranon, S., Rugmai, S. and Daduang, J. (2016). Gallic acid conjugated with gold nanoparticles: antibacterial activity and mechanism of action on foodborne pathogens. *International journal of nanomedicine*, 11: 3347].

[34] Kong, F. Y., Zhang, J. W., Li, R. F., Wang, Z. X., Wang, W. J. and Wang, W. (2017). Unique roles of gold nanoparticles in drug delivery, targeting and imaging applications. *Molecules*, 22 (9): 1445].

[35] Ahmad, T., Wani, I. A., Manzoor, N., Ahmed, J. and Asiri, A. M. (2013). Biosynthesis, structural characterization and antimicrobial activity of gold and silver nanoparticles. *Colloids and Surfaces B: Biointerfaces*, 107: 227-234].

[36] Camas, M., Camas, A. S. and Kyeremeh, K. (2018). Extracellular synthesis and characterization of gold nanoparticles using *Mycobacterium* sp. BRS2A-AR2 isolated from the aerial roots of the Ghanaian mangrove plant, *Rhizophora racemosa*. *Indian journal of microbiology*, 58 (2): 214-221].

[37] Huo, Y., Singh, P., Kim, Y. J., Soshnikova, V., Kang, J., Markus, J., Ahn, S., Castro-Aceituno, V., Mathiyalagan, R. and Chokkalingam, M. (2018). Biological synthesis of gold and silver chloride

nanoparticles by *Glycyrrhiza uralensis* and in vitro applications. *Artificial cells, nanomedicine, and biotechnology*, 46 (2): 303-312].

[38] Maruyama, T., Fujimoto, Y. and Maekawa, T. (2015). Synthesis of gold nanoparticles using various amino acids. *Journal of colloid and interface science*, 447: 254-257].

[39] Wacławek, S., Gončuková, Z., Adach, K., Fijałkowski, M. and Černík, M. (2018). Green synthesis of gold nanoparticles using *Artemisia dracunculus* extract: control of the shape and size by varying synthesis conditions. *Environmental Science and Pollution Research*, 25 (24): 24210-24219].

[40] Geethalakshmi, R. and Sarada, D. (2010). Synthesis of plant-mediated silver nanoparticles using *Trianthema decandra* extract and evaluation of their anti microbial activities. *International Journal of Engineering Science and Technology*, 2 (5): 970-975].

[41] Rajasekharreddy, P., Rani, P. U. and Sreedhar, B. (2010). Qualitative assessment of silver and gold nanoparticle synthesis in various plants: a photobiological approach. *Journal of Nanoparticle Research*, 12 (5): 1711-1721].

[42] Philip, D. and Unni, C. (2011). Extracellular biosynthesis of gold and silver nanoparticles using *Krishna tulsi* (Ocimum sanctum) leaf. *Physica E: Low-dimensional Systems and Nanostructures*, 43 (7): 1318-1322].

[43] Song, J. Y., Jang, H. K. and Kim, B. S. (2009). Biological synthesis of gold nanoparticles using *Magnolia kobus* and *Diopyros kaki* leaf extracts. *Process Biochemistry*, 44 (10): 1133-1138].

[44] Sujitha, M. V. and Kannan, S. (2013). Green synthesis of gold nanoparticles using Citrus fruits (*Citrus limon*, *Citrus reticulata* and *Citrus sinensis*) aqueous extract and its characterization. *Spectrochimica Acta Part A: Molecular and Biomolecular Spectroscopy*, 102: 15-23].

[45] Ankamwar, B. (2010). Biosynthesis of gold nanoparticles (green-gold) using leaf extract of *Terminalia catappa*. *E-Journal of Chemistry*, 7.

[46] Ghodake, G., Deshpande, N., Lee, Y. and Jin, E. (2010). Pear fruit extract-assisted room-temperature biosynthesis of gold nanoplates. *Colloids and Surfaces B: Biointerfaces*, 75 (2): 584-589].

[47] Dauthal, P. and Mukhopadhyay, M. (2012). Prunus domestica fruit extract-mediated synthesis of gold nanoparticles and its catalytic activity for 4-nitrophenol reduction. *Industrial & engineering chemistry research*, 51 (40): 13014-13020].

[48] Daisy, P. and Saipriya, K. (2012). Biochemical analysis of *Cassia fistula* aqueous extract and phytochemically synthesized gold nanoparticles as hypoglycemic treatment for diabetes mellitus. *International journal of nanomedicine*, 7: 1189].

[49] Gangula, A., Podila, R., Karanam, L., Janardhana, C. and Rao, A. M. (2011). Catalytic reduction of 4-nitrophenol using biogenic gold and silver nanoparticles derived from *Breynia rhamnoides*. *Langmuir*, 27 (24): 15268-15274].

[50] Roy, N., Laskar, R. A., Sk, I., Kumari, D., Ghosh, T. and Begum, N. A. (2011). A detailed study on the antioxidant activity of the stem bark of *Dalbergia sissoo* Roxb., an Indian medicinal plant. *Food chemistry*, 126 (3): 1115-1121].

[51] Jayaseelan, C., Ramkumar, R., Rahuman, A. A. and Perumal, P. (2013). Green synthesis of gold nanoparticles using seed aqueous extract of *Abelmoschus esculentus* and its antifungal activity. *Industrial Crops and Products*, 45: 423-429].

[52] Niemeyer, C. M. (2002). The developments of semisynthetic DNA–protein conjugates. *Trends in biotechnology*, 20 (9): 395-401].

[53] Li, Y., Tang, Z., Prasad, P. N., Knecht, M. R. and Swihart, M. T. (2014). Peptide-mediated synthesis of gold nanoparticles: effects of peptide sequence and nature of binding on physicochemical properties. *Nanoscale*, 6 (6): 3165-3172].

[54] Niemeyer, C. M. and Ceyhan, B. (2001). DNA-directed functionalization of colloidal gold with proteins. *Angewandte Chemie International Edition*, 40 (19): 3685-3688].

[55] Liu, J. and Lu, Y. (2006). Preparation of aptamer-linked gold nanoparticle purple aggregates for colorimetric sensing of analytes. *Nature protocols*, 1 (1): 246-252].

[56] Kalaiarasi, R., Jayallakshmi, N. and Venkatachalam, P. (2010). Phytosynthesis of nanoparticles and its applications. *Plant Cell Biotechnology and Molecular Biology*, 11 (1/4): 1-16].

[57] Ebrahiminezhad, A., Zare-Hoseinabadi, A., Sarmah, A. K., Taghizadeh, S., Ghasemi, Y. and Berenjian, A. (2018). Plant-mediated synthesis and applications of iron nanoparticles. *Molecular biotechnology*, 60 (2): 154-168].

[58] Dinali, R., Ebrahiminezhad, A., Manley-Harris, M., Ghasemi, Y. and Berenjian, A. (2017). Iron oxide nanoparticles in modern microbiology and biotechnology. *Critical reviews in microbiology*, 43 (4): 493-507].

[59] Senthil, M. and Ramesh, C. (2012). Biogenic Synthesis of Fe_3O_4 Nanoparticles Using *Tridax procumbens* Leaf Extract and Its Antibacterial Activity on Pseudomonas Aeruginosa. *Digest Journal of Nanomaterials & Biostructures (DJNB)*, 7 (4).

[60] Madhavi, V., Prasad, T., Reddy, A. V. B., Reddy, B. R. and Madhavi, G. (2013). Application of phytogenic zerovalent iron nanoparticles in the adsorption of hexavalent chromium. *Spectrochimica Acta Part A: Molecular and Biomolecular Spectroscopy*, 116: 17-25].

[61] Daniel, S. K., Vinothini, G., Subramanian, N., Nehru, K. and Sivakumar, M. (2013). Biosynthesis of Cu, ZVI, and Ag nanoparticles using *Dodonaea viscosa* extract for antibacterial activity against human pathogens. *Journal of nanoparticle research*, 15 (1): 1319].

[62] Wang, Z., Fang, C. and Megharaj, M. (2014). Characterization of iron–polyphenol nanoparticles synthesized by three plant extracts and their fenton oxidation of azo dye. *ACS Sustainable Chemistry & Engineering*, 2 (4): 1022-1025].

[63] Madduluri, S., Rao, K. B. and Sitaram, B. (2013). In vitro evaluation of antibacterial activity of five indigenous plants extract against five

bacterial pathogens of human. *International Journal of Pharmacy and Pharmaceutical Sciences*, 5 (4): 679-684].

[64] Barua, S., Thakur, S., Aidew, L., Buragohain, A. K., Chattopadhyay, P. and Karak, N. (2014). One step preparation of a biocompatible, antimicrobial reduced graphene oxide–silver nanohybrid as a topical antimicrobial agent. *RSC Advances*, 4 (19): 9777-9783].

[65] Harshiny, M., Iswarya, C. N. and Matheswaran, M. (2015). Biogenic synthesis of iron nanoparticles using *Amaranthus dubius* leaf extract as a reducing agent. *Powder technology*, 286: 744-749].

[66] Martínez-Cabanas, M., López-García, M., Barriada, J. L., Herrero, R. and de Vicente, M. E. S. (2016). Green synthesis of iron oxide nanoparticles. Development of magnetic hybrid materials for efficient As (V) removal. *Chemical Engineering Journal*, 301: 83-91].

[67] Bizuayehu, D., Atlabachew, M. and Ali, M. T. (2016). *Determination of some selected secondary metabolites and their invitro antioxidant activity in commercially available Ethiopian tea (Camellia sinensis)*. SpringerPlus, 5 (1): 412].

[68] Vinodkumar, S., Nakkeeran, S., Renukadevi, P. and Mohankumar, S. (2018). Diversity and antiviral potential of rhizospheric and endophytic *Bacillus* species and phyto-antiviral principles against tobacco streak virus in cotton. *Agriculture, Ecosystems & Environment*, 267: 42-51].

[69] Kumar, B., Smita, K., Cumbal, L., Debut, A., Camacho, J., Hernández-Gallegos, E., Chavez-Lopez, M., Grijalva, M., Angulo, Y. and Rosero, G. (2015). Pomosynthesis and biological activity of silver nanoparticles using *Passiflora tripartita* fruit extracts. *Advanced Materials Letters*, 6 (2): 127-32].

[70] Reddy, N. J., Vali, D. N., Rani, M. and Rani, S. S. (2014). Evaluation of antioxidant, antibacterial and cytotoxic effects of green synthesized silver nanoparticles by *Piper longum* fruit. *Materials Science and Engineering: C*, 34: 115-122].

[71] Venkateswarlu, S., Kumar, B. N., Prasad, C., Venkateswarlu, P. and Jyothi, N. (2014). Bio-inspired green synthesis of Fe_3O_4 spherical

magnetic nanoparticles using *Syzygium cumini* seed extract. *Physica B: Condensed Matter*, 449: 67-71].

[72] Murthy, H. A., Abebe, B., Prakash, C. and Shantaveerayya, K. (2018). A review on green synthesis of Cu and CuO nanomaterials for multifunctional applications. *Material Science Research India*, 15 (3): 279-295].

[73] Marchiol, L. (2012). Synthesis of metal nanoparticles in living plants. *Italian Journal of Agronomy*: e37-e37].

[74] Gültekin, D. D., Güngör, A. A., Önem, H., Babagil, A. and Nadaroğlu, H. (2016). Synthesis of copper nanoparticles using a different method: determination of its antioxidant and antimicrobial activity. *Journal of the Turkish Chemical Society, Section A: Chemistry*, 3 (3): 623-636].

[75] Singh, J., Kumar, V., Kim, K. H. and Rawat, M. (2019). Biogenic synthesis of copper oxide nanoparticles using plant extract and its prodigious potential for photocatalytic degradation of dyes. *Environmental research*, 177: 108569].

[76] Elisma, N., Labanni, A., Rilda, Y., Arief, S. And Asrofi, M. (2019). Green Synthesis of Copper Nanoparticles Using *Uncaria gambir* Roxb. Leaf Extract and Its Characterization. *Rasayan Journal of Chemistry*, 12 (4): 1752-1756].

[77] Kaur, P., Thakur, R. and Chaudhury, A. (2016). Biogenesis of copper nanoparticles using peel extract of *Punica granatum* and their antimicrobial activity against opportunistic pathogens. *Green chemistry letters and reviews*, 9 (1): 33-38].

[78] Ismail, M. (2020). Green synthesis and characterizations of copper nanoparticles. *Materials Chemistry and Physics*, 240: 122283].

[79] Kiranmai, M., Kadimcharla, K., Keesara, N. R., Fatima, S. N., Bommena, P. and Batchu, U. R. (2017). Green synthesis of stable copper nanoparticles and synergistic activity with antibiotics. *Indian journal of pharmaceutical sciences*, 79 (5): 695-700].

[80] Sukumar, S., Rudrasenan, A. and Padmanabhan Nambiar, D. (2020). Green-Synthesized Rice-Shaped Copper Oxide Nanoparticles Using

Caesalpinia bonducella Seed Extract and Their Applications. *ACS omega*, 5 (2): 1040-1051].

[81] Chung, I. M., Abdul Rahuman, A., Marimuthu, S., Vishnu Kirthi, A., Anbarasan, K., Padmini, P. and Rajakumar, G. (2017). Green synthesis of copper nanoparticles using *Eclipta prostrata* leaves extract and their antioxidant and cytotoxic activities. *Experimental and therapeutic medicine*, 14 (1): 18-24].

[82] Ebrahimi, K., Shiravand, S. and Mahmoudvand, H. (2017). Biosynthesis of copper nanoparticles using aqueous extract of *Capparis spinosa* fruit and investigation of its antibacterial activity. *Marmara Pharmaceutical Journal*, 21 (4): 866-871].

[83] Ismail, M., Gul, S., Khan, M., Khan, M. A., Asiri, A. M. and Khan, S. B. (2019). Green synthesis of zerovalent copper nanoparticles for efficient reduction of toxic azo dyes congo red and methyl orange. *Green processing and synthesis*, 8 (1): 135-143].

[84] El-Sheekh, M. M. and El-Kassas, H. Y. (2016). Algal production of nano-silver and gold: Their antimicrobial and cytotoxic activities: A review. *Journal of Genetic Engineering and Biotechnology*, 14 (2): 299-310].

[85] Nirmala, M. J., Shiny, P., Ernest, V., Das, S., Samundeeswari, A., Mukherjee, A. and Chndrasekaran, N. (2013). A review on safer means of nanoparticle synthesis by exploring the prolific marine ecosystem as a new thrust area in nanopharmaceutics. *Int. J. Pharm. Sci*, 5: 23-29].

[86] Sahayaraj, K., Rajesh, S. and Rathi, J. (2012). Silver Nanoparticles Biosynthesis Using Marine Alga *Padina pavonica* (Linn.) and Its Microbicidal Activity. *Digest Journal of Nanomaterials & Biostructures (DJNB)*, 7 (4).

[87] Bhuyar, P., Yusoff, M. M., Rahim, M. H. A., Sundararaju, S., Maniam, G. P. and Govindan, N. (2020). Effect of Plant Hormones on the Production of Biomass and Lipid Extraction for Biodiesel Production from Microalgae *Chlorella* Sp. *Journal of Microbiology, Biotechnology and Food Sciences*, 9 (4): 671-674].

[88] Abdel-Raouf, N., Al-Enazi, N. M., Ibraheem, I. B. M., Alharbi, R. M. and Alkhulaifi, M. M. (2019). Biosynthesis of silver nanoparticles by using of the marine brown alga *Padina pavonia* and their characterization. *Saudi Journal of Biological Sciences*, 26 (6): 1207-1215].

[89] Gajbhiye, S. R., Deshmukh, G. P., Vinu, A. and Kantam, M. L. (2019). Production of p-cymene by Alkylation of Toluene with Propan-2-ol. *Catalysis in Green Chemistry and Engineering*, 2 (2).

[90] Elechiguerra, J. L., Burt, J. L., Morones, J. R., Camacho-Bragado, A., Gao, X., Lara, H. H. and Yacaman, M. J. (2005). Interaction of silver nanoparticles with HIV-1. *Journal of nanobiotechnology*, 3 (1): 1-10].

[91] Chen, L., Xu, S. and Li, J. (2011). Recent advances in molecular imprinting technology: current status, challenges and highlighted applications. *Chemical Society Reviews*, 40 (5): 2922-2942].

[92] Kulkarni, N. and Muddapur, U. (2014). Biosynthesis of metal nanoparticles: a review. *Journal of Nanotechnology*, 2014.

[93] Chisti, Y. 2010. *Fuels from microalgae.* Taylor & Francis.

[94] Luangpipat, T., Beattie, I. R., Chisti, Y. and Haverkamp, R. G. (2011). Gold nanoparticles produced in a microalga. *Journal of Nanoparticle Research*, 13 (12): 6439-6445].

[95] Jegadeeswaran, P., Shivaraj, R. and Venckatesh, R. (2012). Green synthesis of silver nanoparticles from extract of *Padina tetrastromatica* leaf. *Digest Journal of Nanomaterials and Biostructures*, 7 (3): 991-998].

[96] Patil, V. and Sastry, M. (1997). Electrostatically controlled diffusion of carboxylic acid derivatized Q-state CdS nanoparticles in thermally evaporated fatty amine films. *Journal of the Chemical Society, Faraday Transactions*, 93 (24): 4347-4353].

[97] Bhuyar, P., Rahim, M. H. A., Sundararaju, S., Ramaraj, R., Maniam, G. P. and Govindan, N. (2020). Synthesis of silver nanoparticles using marine macroalgae *Padina* sp. and its antibacterial activity towards pathogenic bacteria. *Beni-Suef University Journal of Basic and Applied Sciences*, 9 (1): 1-15].

[98] Brayner, R., Barberousse, H., Hemadi, M., Djedjat, C., Yéprémian, C., Coradin, T., Livage, J., Fiévet, F. and Couté, A. (2007). Cyanobacteria as bioreactors for the synthesis of Au, Ag, Pd, and Pt nanoparticles via an enzyme-mediated route. *Journal of nanoscience and nanotechnology*, 7 (8): 2696-2708].

[99] Lengke, M. F., Fleet, M. E. and Southam, G. (2007). Biosynthesis of silver nanoparticles by filamentous cyanobacteria from a silver (I) nitrate complex. *Langmuir*, 23 (5): 2694-2699]

[100] Aboelfetoh, E. F., El-Shenody, R. A. and Ghobara, M. M. (2017). Eco-friendly synthesis of silver nanoparticles using green algae (*Caulerpa serrulata*): reaction optimization, catalytic and antibacterial activities. *Environmental monitoring and assessment*, 189 (7): 349].

[101] Arya, A., Gupta, K., Chundawat, T. S. and Vaya, D. (2018). Biogenic synthesis of copper and silver nanoparticles using green alga *Botryococcus braunii* and its antimicrobial activity. *Bioinorganic chemistry and applications*, 2018.

[102] Ashokkumar, S., Ravi, S., Kathiravan, V. and Velmurugan, S. 2015. *Retracted: Synthesis of silver nanoparticles using A. indicum leaf extract and their antibacterial activity*. Elsevier.

[103] Barwal, I., Ranjan, P., Kateriya, S. and Yadav, S. C. (2011). Cellular oxido-reductive proteins of *Chlamydomonas reinhardtii* control the biosynthesis of silver nanoparticles. *Journal of nanobiotechnology*, 9 (1): 56].

[104] Jena, J., Pradhan, N., Dash, B. P., Sukla, L. B. and Panda, P. K. (2013). Biosynthesis and characterization of silver nanoparticles using microalga *Chlorococcum humicola* and its antibacterial activity. *Int J Nanomater Biostruct*, 3 (1): 1-8].

[105] El-Sheekh, M. M. and El Kassas, H. Y. (2014). Biosynthesis, characterization and synergistic effect of phytogenic gold nanoparticles by marine picoeukaryote *Picochlorum* sp. in combination with antimicrobials. *Rendiconti Lincei*, 25 (4): 513-521].

[106] Xie, J., Lee, J. Y., Wang, D. I. and Ting, Y. P. (2007). Silver nanoplates: from biological to biomimetic synthesis. *ACS nano*, 1 (5): 429-439].

[107] Moghaddam, K. M. (2010). Department of Pharmaceutical Biotechnology and Student's Scientific Research Center Faculty of Pharmacy, Tehran University of Medical Sciences, Tehran, Iran. *Journal of Young Investigators*.

[108] Rahimi, Z., Yousefzadi, M., Noori, A. and Akbarzadeh, A. (2014). Green synthesis of silver nanoparticles using *Ulva flexousa* from the Persian Gulf, Iran. *Journal of the Persian Gulf*, 5 (15): 9-16].

[109] Vivek, M., Kumar, P. S., Steffi, S. and Sudha, S. (2011). Biogenic silver nanoparticles by *Gelidiella acerosa* extract and their antifungal effects. *Avicenna Journal of Medical Biotechnology*, 3 (3): 143].

[110] Kumar, P., Senthamil Selvi, S., Lakshmi Prabha, A., Prem Kumar, K., Ganeshkumar, R. and Govindaraju, M. (2012). Synthesis of silver nanoparticles from S*argassum tenerrimum* and screening phytochemicals for its antibacterial activity. *Nano Biomed Eng*, 4 (1): 12-16].

[111] Sinha, S. N., Paul, D., Halder, N., Sengupta, D. and Patra, S. K. (2015). Green synthesis of silver nanoparticles using fresh water green alga *Pithophora oedogonia* (Mont.) Wittrock and evaluation of their antibacterial activity. *Applied Nanoscience*, 5 (6): 703-709].

[112] Ahmadi, F. S., Tanhaeian, A. and Pirkohi, M. H. (2016). Biosynthesis of silver nanoparticles using *Chlamydomonas reinhardtii* and its inhibitory effect on growth and virulence of Listeria monocytogenes. *Iranian Journal of Biotechnology*, 14 (3): 163].

[113] Shende, S., Gade, A. and Rai, M. (2017). Large-scale synthesis and antibacterial activity of fungal-derived silver nanoparticles. *Environmental Chemistry Letters*, 15 (3): 427-434].

[114] Singaravelu, G., Arockiamary, J., Kumar, V. G. and Govindaraju, K. (2007). A novel extracellular synthesis of monodisperse gold

nanoparticles using marine alga, *Sargassum wightii* Greville. *Colloids and surfaces B: Biointerfaces*, 57 (1): 97-101].

[115] Iravani, S., Korbekandi, H., Mirmohammadi, S. V. and Zolfaghari, B. (2014). Synthesis of silver nanoparticles: chemical, physical and biological methods. *Research in pharmaceutical sciences*, 9 (6): 385].

[116] El-Kassas, H. Y. and El-Sheekh, M. M. (2014). Cytotoxic activity of biosynthesized gold nanoparticles with an extract of the red seaweed *Corallina officinalis* on the MCF-7 human breast cancer cell line. *Asian Pacific journal of cancer prevention*, 15 (10): 4311-4317].

[117] Sehgal, N., Soni, K., Gupta, N. and Kanchan, K. (2018). Microorganism assisted synthesis of gold nanoparticles: a review. *Asian J Biomed Pharm Sci*, 8 (64): 22-29].

[118] Rajeshkumar, S., Malarkodi, C., Vanaja, M., Gnanajobitha, G., Paulkumar, K., Kannan, C. and Annadurai, G. (2013). Antibacterial activity of algae mediated synthesis of gold nanoparticles from *Turbinaria conoides*. *Der Pharma Chemica*, 5 (2): 224-229].

[119] Senthilkumar, P., Surendran, L., Sudhagar, B. and Kumar, D. R. S. (2019). Facile green synthesis of gold nanoparticles from marine algae *Gelidiella acerosa* and evaluation of its biological Potential. *SN Applied Sciences*, 1 (4): 284].

[120] Lengke, M. F., Fleet, M. E. and Southam, G. (2006). Morphology of gold nanoparticles synthesized by filamentous cyanobacteria from gold (I)− thiosulfate and gold (III)− chloride complexes. *Langmuir*, 22 (6): 2780-2787].

[121] Parial, D., Patra, H. K., Roychoudhury, P., Dasgupta, A. K. and Pal, R. (2012). Gold nanorod production by cyanobacteria—a green chemistry approach. *Journal of applied phycology*, 24 (1): 55-60].

[122] Parial, D., Gopal, P. K., Paul, S. and Pal, R. (2016). Gold (III) bioreduction by cyanobacteria with special reference to in vitro biosafety assay of gold nanoparticles. *Journal of applied phycology*, 28 (6): 3395-3406].

[123] Xie, J., Lee, J. Y. and Wang, D. I. (2007). Synthesis of single-crystalline gold nanoplates in aqueous solutions through

biomineralization by serum albumin protein. *The Journal of Physical Chemistry C*, 111 (28): 10226-10232].

[124] Rajathi, F. A. A., Parthiban, C., Kumar, V. G. and Anantharaman, P. (2012). Biosynthesis of antibacterial gold nanoparticles using brown alga, *Stoechospermum marginatum* (kützing). *Spectrochimica Acta Part A: Molecular and Biomolecular Spectroscopy*, 99: 166-173].

[125] Naveena, B. E. and Prakash, S. (2013). Biological synthesis of gold nanoparticles using marine algae *Gracilaria corticata* and its application as a potent antimicrobial and antioxidant agent. *Asian J Pharm Clin Res*, 6 (2): 179-182].

[126] Rajeshkumar, S., Malarkodi, C., Gnanajobitha, G., Paulkumar, K., Vanaja, M., Kannan, C. and Annadurai, G. (2013). Seaweed-mediated synthesis of gold nanoparticles using *Turbinaria conoides* and its characterization. *Journal of Nanostructure in Chemistry*, 3 (1): 44].

[127] Chen, J., Han, H., Chen, M., Xu, X. Z., Wang, B. and Shi, L. Y. (2014). Inactivated Sendai virus strain Tianjin induces apoptosis in human breast cancer MDA-MB-231 cells. *Asian Pac J Cancer Prev*, 15 (12): 5023-5028].

[128] Kayalvizhi, K., Asmathunisha, N., Subramanian, V. and Kathiresan, K. (2014). Purification of silver and gold nanoparticles from two species of brown seaweeds (Padina tetrastromatica and Turbinaria ornata). *J. Med. Plants Stud*, 2 (4): 32-37].

[129] Namvar, F., Mohamed, S., Fard, S. G., Behravan, J., Mustapha, N. M., Alitheen, N. B. M. and Othman, F. (2012). Polyphenol-rich seaweed (*Eucheuma cottonii*) extract suppresses breast tumour via hormone modulation and apoptosis induction. *Food chemistry*, 130 (2): 376-382].

[130] Zuercher, A. W., Fritsche, R., Corthésy, B. and Mercenier, A. (2006). Food products and allergy development, prevention and treatment. *Current opinion in biotechnology*, 17 (2): 198-203].

[131] Miyashita, K. (2009). The carotenoid fucoxanthin from brown seaweed affects obesity. *Lipid Technology*, 21 (8-9): 186-190].

[132] Mohamed, S., Hashim, S. N. and Rahman, H. A. (2012). Seaweeds: a sustainable functional food for complementary and alternative therapy. *Trends in Food Science & Technology*, 23 (2): 83-96].

[133] Wada, K., Nakamura, K., Tamai, Y., Tsuji, M., Sahashi, Y., Watanabe, K., Ohtsuchi, S., Yamamoto, K., Ando, K. and Nagata, C. (2011). Seaweed intake and blood pressure levels in healthy pre-school Japanese children. *Nutrition Journal*, 10 (1): 83].

[134] Mahdavi, M., Namvar, F., Ahmad, M. B. and Mohamad, R. (2013). Green biosynthesis and characterization of magnetic iron oxide (Fe3O4) nanoparticles using seaweed (*Sargassum muticum*) aqueous extract. *Molecules*, 18 (5): 5954-5964].

[135] Roy, S. *A Review: Green Synthesis of Nanoparticles from Seaweeds and Its some Applications*.

[136] Venkatpurwar, V. (2011). Pokharkar V. Green synthesis of silver nanoparticles by *Chrysanthemum morifolium* Ramat. extract and their application in clinical ultrasound gel. *Mater Lett*, 65: 999-1002].

[137] Mahdavi, M., Ahmad, M. B., Haron, M. J., Gharayebi, Y., Shameli, K. and Nadi, B. (2013). Fabrication and characterization of SiO 2/(3-aminopropyl) triethoxysilane-coated magnetite nanoparticles for lead (II) removal from aqueous solution. *Journal of Inorganic and Organometallic Polymers and Materials*, 23 (3): 599-607].

[138] Salem, D. M., Ismail, M. M. and Aly-Eldeen, M. A. (2019). Biogenic synthesis and antimicrobial potency of iron oxide (Fe3O4) nanoparticles using algae harvested from the Mediterranean Sea, Egypt. *The Egyptian Journal of Aquatic Research*, 45 (3): 197-204].

[139] El-Kassas, H. Y., Aly-Eldeen, M. A. and Gharib, S. M. (2016). Green synthesis of iron oxide (Fe 3 O 4) nanoparticles using two selected brown seaweeds: characterization and application for lead bioremediation. *Acta Oceanologica Sinica*, 35 (8): 89-98].

[140] Siji, S., Njana, J., Amrita, P. and Vishnudasan, D. (2018). Biogenic synthesis of iron oxide nanoparticles from marine algae. *TKM International Journal for Multidisciplinary Research*, 1 (1): 1-7].

[141] Swamy, M. K., Sudipta, K., Jayanta, K. and Balasubramanya, S. (2015). The green synthesis, characterization, and evaluation of the biological activities of silver nanoparticles synthesized from *Leptadenia reticulata* leaf extract. *Applied nanoscience*, 5 (1): 73-81].

[142] Noguez, C. (2007). Surface plasmons on metal nanoparticles: the influence of shape and physical environment. *The Journal of Physical Chemistry C*, 111 (10): 3806-3819].

[143] Abboud, Y., Saffaj, T., Chagraoui, A., El Bouari, A., Brouzi, K., Tanane, O. and Ihssane, B. (2014). Biosynthesis, characterization and antimicrobial activity of copper oxide nanoparticles (CONPs) produced using brown alga extract (*Bifurcaria bifurcata*). *Applied Nanoscience*, 4 (5): 571-576].

[144] Borgohain, K., Murase, N. and Mahamuni, S. (2002). Synthesis and properties of Cu 2 O quantum particles. *Journal of applied physics*, 92 (3): 1292-1297].

[145] Kimber, R. L., Lewis, E. A., Parmeggiani, F., Smith, K., Bagshaw, H., Starborg, T., Joshi, N., Figueroa, A. I., van der Laan, G. and Cibin, G. (2018). Biosynthesis and characterization of copper nanoparticles using *Shewanella oneidensis*: application for click chemistry. *Small*, 14 (10): 1703145].

[146] Yin, M., Wu, C. K., Lou, Y., Burda, C., Koberstein, J. T., Zhu, Y. and O'Brien, S. (2005). Copper oxide nanocrystals. *Journal of the American Chemical Society*, 127 (26): 9506-9511].

[147] Krithiga, N., Jayachitra, A. and Rajalakshmi, A. (2013). Synthesis, characterization and analysis of the effect of copper oxide nanoparticles in biological systems. *Ind J Ns*, 1: 6-15].

[148] Ravindra, B. and Rajasab, A. (2014). *A comparative study on biosynthesis of silver nanoparticles using four different fungal species.*

[149] Rahman, A., Ismail, A., Jumbianti, D., Magdalena, S. and Sudrajat, H. (2009). Synthesis of copper oxide nano particles by using *Phormidium cyanobacterium. Indonesian Journal of Chemistry*, 9 (3): 355-360].

[150] Volanti, D., Keyson, D., Cavalcante, L., Simões, A. Z., Joya, M., Longo, E., Varela, J. A., Pizani, P. and Souza, A. (2008). Synthesis and characterization of CuO flower-nanostructure processing by a domestic hydrothermal microwave. *Journal of Alloys and Compounds*, 459 (1-2): 537-542].

[151] Ramaswamy, S. V. P., Narendhran, S. and Sivaraj, R. (2016). Potentiating effect of ecofriendly synthesis of copper oxide nanoparticles using brown alga: antimicrobial and anticancer activities. *Bulletin of Materials Science*, 39 (2): 361-364].

[152] Saharan, V., Mehrotra, A., Khatik, R., Rawal, P., Sharma, S. and Pal, A. (2013). Synthesis of chitosan based nanoparticles and their in vitro evaluation against phytopathogenic fungi. *International journal of biological macromolecules*, 62: 677-683].

[153] Fayaz, A. M., Balaji, K., Kalaichelvan, P. and Venkatesan, R. (2009). Fungal based synthesis of silver nanoparticles—an effect of temperature on the size of particles. *Colloids and Surfaces B: Biointerfaces*, 74 (1): 123-126].

[154] Ahmad, A., Mukherjee, P., Senapati, S., Mandal, D., Khan, M. I., Kumar, R. and Sastry, M. (2003). Extracellular biosynthesis of silver nanoparticles using the fungus *Fusarium oxysporum*. *Colloids and surfaces B: Biointerfaces*, 28 (4): 313-318].

[155] Jo, Y. K., Kim, B. H. and Jung, G. (2009). Antifungal activity of silver ions and nanoparticles on phytopathogenic fungi. *Plant disease*, 93 (10): 1037-1043].

[156] Siddiqi, K. S. and Husen, A. (2016). Fabrication of metal nanoparticles from fungi and metal salts: scope and application. *Nanoscale research letters*, 11 (1): 98].

[157] Elamawi, R. M., Al-Harbi, R. E. and Hendi, A. A. (2018). Biosynthesis and characterization of silver nanoparticles using *Trichoderma longibrachiatum* and their effect on phytopathogenic fungi. *Egyptian journal of biological pest control*, 28 (1): 28].

[158] Rodríguez-González, V., Obregón, S., Patrón-Soberano, O. A., Terashima, C. and Fujishima, A. (2020). An approach to the photocatalytic mechanism in the TiO_2-nanomaterials microorganism

interface for the control of infectious processes. *Applied Catalysis B: Environmental*: 118853].

[159] Ingle, A., Rai, M., Gade, A. and Bawaskar, M. (2009). Fusarium solani: a novel biological agent for the extracellular synthesis of silver nanoparticles. *Journal of Nanoparticle Research*, 11 (8): 2079].

[160] Syed, A., Saraswati, S., Kundu, G. C. and Ahmad, A. (2013). Biological synthesis of silver nanoparticles using the fungus Humicola sp. and evaluation of their cytoxicity using normal and cancer cell lines. *Spectrochimica Acta Part A: Molecular and Biomolecular Spectroscopy*, 114: 144-147].

[161] Korbekandi, H., Ashari, Z., Iravani, S. and Abbasi, S. (2013). Optimization of biological synthesis of silver nanoparticles using *Fusarium oxysporum*. *Iranian journal of pharmaceutical research: IJPR*, 12 (3): 289].

[162] Owaid, M. N., Raman, J., Lakshmanan, H., Al-Saeedi, S. S. S., Sabaratnam, V. and Abed, I. A. (2015). Mycosynthesis of silver nanoparticles by Pleurotus cornucopiae var. citrinopileatus and its inhibitory effects against *Candida* sp. *Materials Letters*, 153: 186-190].

[163] Al-Bahrani, R., Raman, J., Lakshmanan, H., Hassan, A. A. and Sabaratnam, V. (2017). Green synthesis of silver nanoparticles using tree oyster mushroom Pleurotus ostreatus and its inhibitory activity against pathogenic bacteria. *Materials Letters*, 186: 21-25].

[164] Durán, N., Marcato, P. D., Alves, O. L., De Souza, G. I. and Esposito, E. (2005). Mechanistic aspects of biosynthesis of silver nanoparticles by several *Fusarium oxysporum* strains. *Journal of nanobiotechnology*, 3 (1): 8].

[165] Durán, N., Marcato, P. D., De Souza, G. I., Alves, O. L. and Esposito, E. (2007). Antibacterial effect of silver nanoparticles produced by fungal process on textile fabrics and their effluent treatment. *Journal of biomedical nanotechnology*, 3 (2): 203-208].

[166] Li, G., He, D., Qian, Y., Guan, B., Gao, S., Cui, Y., Yokoyama, K. and Wang, L. (2012). Fungus-mediated green synthesis of silver

nanoparticles using *Aspergillus terreus*. *International journal of molecular sciences*, 13 (1): 466-476].

[167] Vigneshwaran, N., Ashtaputre, N., Varadarajan, P., Nachane, R., Paralikar, K. and Balasubramanya, R. (2007). Biological synthesis of silver nanoparticles using the fungus *Aspergillus flavus*. *Materials letters*, 61 (6): 1413-1418].

[168] Bhainsa, K. C. and D'souza, S. (2006). Extracellular biosynthesis of silver nanoparticles using the fungus *Aspergillus fumigatus*. *Colloids and surfaces B: Biointerfaces*, 47 (2): 160-164].

[169] Kowshik, M., Ashtaputre, S., Kharrazi, S., Vogel, W., Urban, J., Kulkarni, S. K. and Paknikar, K. (2002). Extracellular synthesis of silver nanoparticles by a silver-tolerant yeast strain MKY3. *Nanotechnology*, 14 (1): 95].

[170] Daniel, M. C. and Astruc, D. (2004). Gold nanoparticles: assembly, supramolecular chemistry, quantum-size-related properties, and applications toward biology, catalysis, and nanotechnology. *Chemical reviews*, 104 (1): 293-346].

[171] Murphy, C. J., Gole, A. M., Stone, J. W., Sisco, P. N., Alkilany, A. M., Goldsmith, E. C. and Baxter, S. C. (2008). Gold nanoparticles in biology: beyond toxicity to cellular imaging. *Accounts of chemical research*, 41 (12): 1721-1730].

[172] Andreescu, D., Sau, T. K. and Goia, D. V. (2006). Stabilizer-free nanosized gold sols. *Journal of colloid and interface science*, 298 (2): 742-751].

[173] Shankar, S. S., Ahmad, A., Pasricha, R. and Sastry, M. (2003). Bioreduction of chloroaurate ions by geranium leaves and its endophytic fungus yields gold nanoparticles of different shapes. *Journal of Materials Chemistry*, 13 (7): 1822-1826].

[174] Mishra, A., Tripathy, S. K., Wahab, R., Jeong, S. H., Hwang, I., Yang, Y. B., Kim, Y. S., Shin, H. S. and Yun, S. I. (2011). Microbial synthesis of gold nanoparticles using the fungus *Penicillium brevicompactum* and their cytotoxic effects against mouse mayo blast cancer C 2 C 12 cells. *Applied microbiology and biotechnology*, 92 (3): 617-630].

[175] Gupta, S. and Bector, S. (2013). Biosynthesis of extracellular and intracellular gold nanoparticles by *Aspergillus fumigatus* and *A. flavus*. *Antonie Van Leeuwenhoek*, 103 (5): 1113-1123].

[176] Narayanan, K. B. and Sakthivel, N. (2011). Facile green synthesis of gold nanostructures by NADPH-dependent enzyme from the extract of *Sclerotium rolfsii*. *Colloids and Surfaces A: Physicochemical and Engineering Aspects*, 380 (1-3): 156-161].

[177] Mukherjee, P., Ahmad, A., Mandal, D., Senapati, S., Sainkar, S. R., Khan, M. I., Ramani, R., Parischa, R., Ajayakumar, P. and Alam, M. (2001). Bioreduction of AuCl4− ions by the fungus, *Verticillium* sp. and surface trapping of the gold nanoparticles formed. *Angewandte Chemie International Edition*, 40 (19): 3585-3588].

[178] Vala, A. K. (2015). Exploration on green synthesis of gold nanoparticles by a marine-derived fungus *Aspergillus sydowii*. *Environmental Progress & Sustainable Energy*, 34 (1): 194-197].

[179] Sheikhloo, Z., Salouti, M. and Katiraee, F. (2011). Biological synthesis of gold nanoparticles by fungus *Epicoccumnigrum*. *Journal of Cluster Science*, 22 (4): 661-665].

[180] Sanghi, R., Verma, P. and Puri, S. (2011). Enzymatic formation of gold nanoparticles using *Phanerochaete chrysosporium*. *Advances in Chemical Engineering and Science*, 1 (03): 154].

[181] Kumar, S. A., Peter, Y. A. and Nadeau, J. L. (2008). Facile biosynthesis, separation and conjugation of gold nanoparticles to doxorubicin. *Nanotechnology*, 19 (49): 495101].

[182] Mishra, A., Tripathy, S. K. and Yun, S. I. (2011). Bio-synthesis of gold and silver nanoparticles from *Candida guilliermondii* and their antimicrobial effect against pathogenic bacteria. *Journal of nanoscience and nanotechnology*, 11 (1): 243-248].

[183] Pimprikar, P., Joshi, S., Kumar, A., Zinjarde, S. and Kulkarni, S. (2009). Influence of biomass and gold salt concentration on nanoparticle synthesis by the tropical marine yeast *Yarrowia lipolytica* NCIM 3589. *Colloids and Surfaces B: Biointerfaces*, 74 (1): 309-316].

[184] Clarance, P., Luvankar, B., Sales, J., Khusro, A., Agastian, P., Tack, J. C., Al Khulaifi, M. M., AL-Shwaiman, H. A., Elgorban, A. M. and Syed, A. (2020). Green synthesis and characterization of gold nanoparticles using endophytic fungi *Fusarium solani* and its in-vitro anticancer and biomedical applications. *Saudi Journal of Biological Sciences*, 27 (2): 706-712].

[185] Molnár, Z., Bódai, V., Szakacs, G., Erdélyi, B., Fogarassy, Z., Sáfrán, G., Varga, T., Kónya, Z., Tóth-Szeles, E. and Szűcs, R. (2018). Green synthesis of gold nanoparticles by thermophilic filamentous fungi. *Scientific reports*, 8 (1): 1-12].

[186] Bharde, A., Rautaray, D., Bansal, V., Ahmad, A., Sarkar, I., Yusuf, S. M., Sanyal, M. and Sastry, M. (2006). Extracellular biosynthesis of magnetite using fungi. *Small*, 2 (1): 135-141].

[187] Kaul, R., Kumar, P., Burman, U., Joshi, P., Agrawal, A., Raliya, R. and Tarafdar, J. (2012). Magnesium and iron nanoparticles production using microorganisms and various salts. *Materials Science-Poland*, 30 (3): 254-258].

[188] Mohamed, Y., Azzam, A., Amin, B. and Safwat, N. (2015). Mycosynthesis of iron nanoparticles by *Alternaria alternata* and its antibacterial activity. *African Journal of Biotechnology*, 14 (14): 1234-1241].

[189] Mahanty, S., Bakshi, M., Ghosh, S., Chatterjee, S., Bhattacharyya, S., Das, P., Das, S. and Chaudhuri, P. (2019). Green synthesis of iron oxide nanoparticles mediated by filamentous fungi isolated from Sundarban Mangrove ecosystem, India. *BioNanoScience*, 9 (3): 637-651].

[190] Nasser, F. and Lynch, I. (2016). Secreted protein eco-corona mediates uptake and impacts of polystyrene nanoparticles on *Daphnia magna*. *Journal of proteomics*, 137: 45-51].

[191] Bhargava, A., Jain, N., Barathi, M., Akhtar, M. S., Yun, Y. S. and Panwar, J. 2013. "Synthesis, characterization and mechanistic insights of mycogenic iron oxide nanoparticles." In *Nanotechnology for sustainable development*, 337-348. Springer.

[192] Abdeen, M., Sabry, S., Ghozlan, H., El-Gendy, A. A. and Carpenter, E. E. (2016). Microbial-physical synthesis of Fe and Fe3O4 magnetic nanoparticles using *Aspergillus niger* YESM1 and supercritical condition of ethanol. *Journal of Nanomaterials*, 2016.

[193] Mahanty, S., Bakshi, M., Ghosh, S., Gaine, T., Chatterjee, S., Bhattacharyya, S., Das, S., Das, P. and Chaudhuri, P. (2019). Mycosynthesis of iron oxide nanoparticles using manglicolous fungi isolated from Indian sundarbans and its application for the treatment of chromium containing solution: Synthesis, adsorption isotherm, kinetics and thermodynamics study. *Environmental Nano-technology, Monitoring & Management*, 12: 100276].

[194] Mazumdar, H. and Haloi, N. (2011). A study on biosynthesis of iron nanoparticles by *Pleurotus* sp. *J Microbiol Biotechnol Res*, 1 (3): 39-49].

[195] Cuevas, R., Durán, N., Diez, M., Tortella, G. and Rubilar, O. (2015). Extracellular biosynthesis of copper and copper oxide nanoparticles by *Stereum hirsutum*, a native white-rot fungus from chilean forests. *Journal of Nanomaterials*, 2015.

[196] Honary, S., Barabadi, H., Gharaei-Fathabad, E. and Naghibi, F. (2012). Green synthesis of copper oxide nanoparticles using *Penicillium aurantiogriseum*, *Penicillium citrinum* and *Penicillium waksmanii*. *Dig J Nanomater Bios*, 7 (3): 999-1005].

[197] Hosseini, M., Schaffie, M., Pazouki, M., Darezereshki, E. and Ranjbar, M. (2012). Biologically synthesized copper sulfide nanoparticles: Production and characterization. *Materials Science in Semiconductor Processing*, 15 (2): 222-225].

[198] Noor, S., Shah, Z., Javed, A., Ali, A., Hussain, S. B., Zafar, S., Ali, H. and Muhammad, S. A. (2020). A fungal based synthesis method for copper nanoparticles with the determination of anticancer, antidiabetic and antibacterial activities. *Journal of Microbiological Methods*: 105966].

[199] Logeswari, P., Silambarasan, S. and Abraham, J. (2015). Synthesis of silver nanoparticles using plants extract and analysis of their

antimicrobial property. *Journal of Saudi Chemical Society*, 19 (3): 311-317].

[200] Jain, D., Daima, H. K., Kachhwaha, S. and Kothari, S. (2009). Synthesis of plant-mediated silver nanoparticles using papaya fruit extract and evaluation of their anti microbial activities. *Digest journal of nanomaterials and biostructures*, 4 (3): 557-563].

[201] Gnanadesigan, M., Anand, M., Ravikumar, S., Maruthupandy, M., Ali, M. S., Vijayakumar, V. and Kumaraguru, A. (2012). Antibacterial potential of biosynthesised silver nanoparticles using *Avicennia marina* mangrove plant. *Applied Nanoscience*, 2 (2): 143-147].

[202] Nayak, D., Pradhan, S., Ashe, S., Rauta, P. R. and Nayak, B. (2015). Biologically synthesised silver nanoparticles from three diverse family of plant extracts and their anticancer activity against epidermoid A431 carcinoma. *Journal of colloid and interface science*, 457: 329-338].

[203] Vijilvani, C., Bindhu, M., Frincy, F., AlSalhi, M. S., Sabitha, S., Saravanakumar, K., Devanesan, S., Umadevi, M., Aljaafreh, M. J. and Atif, M. (2020). Antimicrobial and catalytic activities of biosynthesized gold, silver and palladium nanoparticles from *Solanum nigurum* leaves. *Journal of Photochemistry and Photobiology B: Biology*, 202: 111713].

[204] Annamalai, A., Christina, V., Sudha, D., Kalpana, M. and Lakshmi, P. (2013). Green synthesis, characterization and antimicrobial activity of Au NPs using *Euphorbia hirta* L. leaf extract. *Colloids and Surfaces B: Biointerfaces*, 108: 60-65].

[205] Chandran, K., Song, S. and Yun, S. I. (2019). Effect of size and shape controlled biogenic synthesis of gold nanoparticles and their mode of interactions against food borne bacterial pathogens. *Arabian Journal of Chemistry*, 12 (8): 1994-2006].

[206] Aljabali, A. A., Akkam, Y., Al Zoubi, M. S., Al-Batayneh, K. M., Al-Trad, B., Abo Alrob, O., Alkilany, A. M., Benamara, M. and Evans, D. J. (2018). Synthesis of gold nanoparticles using leaf

extract of *Ziziphus zizyphus* and their antimicrobial activity. *Nanomaterials*, 8 (3): 174].

[207] Naseem, T. and Farrukh, M. A. (2015). Antibacterial activity of green synthesis of iron nanoparticles using *Lawsonia inermis* and *Gardenia jasminoides* leaves extract. *Journal of Chemistry*, 2015.

[208] Aziz, W. J. and Abd Urabe, A. (2019). Chemical Preparation of Iron Oxide Nanoparticles Using Plants Extracts in Antibacterial Application. *International Journal of Bioorganic Chemistry*, 4 (1): 1].

[209] Groiss, S., Selvaraj, R., Varadavenkatesan, T. and Vinayagam, R. (2017). Structural characterization, antibacterial and catalytic effect of iron oxide nanoparticles synthesised using the leaf extract of *Cynometra ramiflora*. *Journal of Molecular Structure*, 1128: 572-578].

[210] Soplanit, V. S., Taufiq, A., Hidayat, A., Nikmah, A., Yuliantika, D., Ulya, H. N. and Saputra, R. E. 2020. "Synthesis of Fe3O4/α-Fe2O3/ZnO nanocomposite for antibacterial application." *AIP Conference Proceedings*.

[211] Madubuonu, N., Aisida, S. O., Ali, A., Ahmad, I., Zhao, T. K., Botha, S., Maaza, M. and Ezema, F. I. (2019). Biosynthesis of iron oxide nanoparticles via a composite of *Psidium guavaja-Moringa oleifera* and their antibacterial and photocatalytic study. *Journal of Photochemistry and Photobiology B: Biology*, 199: 111601].

[212] Sathishkumar, G., Logeshwaran, V., Sarathbabu, S., Jha, P. K., Jeyaraj, M., Rajkuberan, C., Senthilkumar, N. and Sivaramakrishnan, S. (2018). Green synthesis of magnetic Fe3O4 nanoparticles using *Couroupita guianensis* Aubl. fruit extract for their antibacterial and cytotoxicity activities. *Artificial cells, nanomedicine, and biotechnology*, 46 (3): 589-598].

[213] Gyawali, R., Ibrahim, S. A., Abu Hasfa, S. H., Smqadri, S. Q. and Haik, Y. (2011). Antimicrobial activity of copper alone and in combination with lactic acid against *Escherichia coli* O157: H7 in laboratory medium and on the surface of lettuce and tomatoes. *Journal of pathogens*, 2011.

[214] Grass, G., Rensing, C. and Solioz, M. (2011). Metallic copper as an antimicrobial surface. *Applied and environmental microbiology*, 77 (5): 1541-1547].

[215] Raffi, M., Mehrwan, S., Bhatti, T. M., Akhter, J. I., Hameed, A., Yawar, W. and ul Hasan, M. M. (2010). Investigations into the antibacterial behavior of copper nanoparticles against *Escherichia coli*. *Annals of microbiology*, 60 (1): 75-80].

[216] Chatterjee, A. K., Sarkar, R. K., Chattopadhyay, A. P., Aich, P., Chakraborty, R. and Basu, T. (2012). A simple robust method for synthesis of metallic copper nanoparticles of high antibacterial potency against E. coli. *Nanotechnology*, 23 (8): 085103].

[217] Sutradhar, P., Saha, M. and Maiti, D. (2014). Microwave synthesis of copper oxide nanoparticles using tea leaf and coffee powder extracts and its antibacterial activity. *Journal of Nanostructure in Chemistry*, 4 (1): 86].

[218] Gurunathan, S., Jeong, J. K., Han, J. W., Zhang, X. F., Park, J. H. and Kim, J. H. (2015). Multidimensional effects of biologically synthesized silver nanoparticles in *Helicobacter pylori*, *Helicobacter felis*, and human lung (L132) and lung carcinoma A549 cells. *Nanoscale research letters*, 10 (1): 1-17].

[219] Morones, J. R., Elechiguerra, J. L., Camacho, A., Holt, K., Kouri, J. B., Ramírez, J. T. and Yacaman, M. J. (2005). The bactericidal effect of silver nanoparticles. *Nanotechnology*, 16 (10): 2346].

[220] Rajesh, S., Raja, D. P., Rathi, J. and Sahayaraj, K. (2012). Biosynthesis of silver nanoparticles using *Ulva fasciata* (Delile) ethyl acetate extract and its activity against Xanthomonas campestris pv. malvacearum. *Journal of Biopesticides*, 5: 119].

[221] Carlson, C., Hussain, S. M., Schrand, A. M., K. Braydich-Stolle, L., Hess, K. L., Jones, R. L. and Schlager, J. J. (2008). Unique cellular interaction of silver nanoparticles: size-dependent generation of reactive oxygen species. *The journal of physical chemistry B*, 112 (43): 13608-13619].

[222] Devi, J. S. and Bhimba, B. V. (2012). Silver nanoparticles: Antibacterial activity against wound isolates & invitro cytotoxic

activity on Human Caucasian colon adenocarcinoma. *Asian Pacific Journal of Tropical Disease*, 2: S87-S93].

[223] Merin, D. D., Prakash, S. and Bhimba, B. V. (2010). Antibacterial screening of silver nanoparticles synthesized by marine micro algae. *Asian Pacific Journal of Tropical Medicine*, 3 (10): 797-799].

[224] Suriya, J., Raja, S. B., Sekar, V. and Rajasekaran, R. (2012). Biosynthesis of silver nanoparticles and its antibacterial activity using seaweed *Urospora* sp. *African Journal of Biotechnology*, 11 (58): 12192-12198].

[225] Roy, S. and Anantharaman, P. (2018). Biosynthesis of Silver Nanoparticles by Sargassum Ilicifolium (Turner) C. *Agardh with their Antimicrobial Activity and Potential for Seed Germination*: 2].

[226] Slavin, Y. N., Asnis, J., Häfeli, U. O. and Bach, H. (2017). Metal nanoparticles: understanding the mechanisms behind antibacterial activity. *Journal of nanobiotechnology*, 15 (1): 1-20].

[227] Darroudi, M., Ahmad, M. and Mashreghi, M. (2014). Gelatinous silver colloid nanoparticles: Synthesis, characterization, and their antibacterial activity. *Journal of Optoelectronics and Advanced Materials*, 16 (1): 182-187].

[228] Guzmán, M. G., Dille, J. and Godet, S. (2009). Synthesis of silver nanoparticles by chemical reduction method and their antibacterial activity. *Int J Chem Biomol Eng*, 2 (3): 104-111].

[229] Lee, G., Lee, H., Nam, K., Han, J. H., Yang, J., Lee, S. W., Yoon, D. S., Eom, K. and Kwon, T. (2012). Nanomechanical characterization of chemical interaction between gold nanoparticles and chemical functional groups. *Nanoscale research letters*, 7 (1): 608].

[230] Subramaniyam, V., Subashchandrabose, S. R., Thavamani, P., Megharaj, M., Chen, Z. and Naidu, R. (2015). *Chlorococcum* sp. MM11—a novel phyco-nanofactory for the synthesis of iron nanoparticles. *Journal of Applied Phycology*, 27 (5): 1861-1869].

[231] Prakash, S., Elavarasan, N., Venkatesan, A., Subashini, K., Sowndharya, M. and Sujatha, V. (2018). Green synthesis of copper oxide nanoparticles and its effective applications in Biginelli

reaction, BTB photodegradation and antibacterial activity. *Advanced Powder Technology*, 29 (12): 3315-3326].

[232] Zhang, H. and Chen, G. (2009). Potent antibacterial activities of Ag/TiO2 nanocomposite powders synthesized by a one-pot sol− gel method. *Environmental science & technology*, 43 (8): 2905-2910].

[233] Baker, C., Pradhan, A., Pakstis, L., Pochan, D. J. and Shah, S. I. (2005). Synthesis and antibacterial properties of silver nanoparticles. *Journal of nanoscience and nanotechnology*, 5 (2): 244-249].

[234] Chandran, M., Yuvaraj, D., Christudhas, L. and Ramesh, K. (2016). Bio synthesis of iron nanoparticles using the brown seaweed, *Dictyota dicotoma*. *Biotechnol. Indian J*, 12 (12): 112]

[235] Rana, S. and Kalaichelvan, P. (2011). Antibacterial activities of metal nanoparticles. *Antibacterial Activities of Metal Nanoparticles*, 11 (02): 21-23].

[236] Saliba, A. M., De Assis, M. C., Nishi, R., Raymond, B., Marques, E. d. A., Lopes, U. G., Touqui, L. and Plotkowski, M. C. (2006). Implications of oxidative stress in the cytotoxicity of *Pseudomonas aeruginosa* ExoU. *Microbes and infection*, 8 (2): 450-459].

[237] Rezaei, Z. S., Javed, A., Ghani, M. J., Soufian, S., Barzegari, F. F., Bayandori, M. A. and Mirjalili, S. H. (2010). *Comparative study of antimicrobial activities of* TiO_2 *and CdO nanoparticles against the pathogenic strain of Escherichia coli*.

[238] Eom, S. H., Kim, Y. M. and Kim, S. K. (2012). Antimicrobial effect of phlorotannins from marine brown algae. *Food and Chemical Toxicology*, 50 (9): 3251-3255].

[239] Stohs, S. J. and Bagchi, D. (1995). Oxidative mechanisms in the toxicity of metal ions. *Free radical biology and medicine*, 18 (2): 321-336].

[240] Beveridge, T. and Murray, R. (1980). Sites of metal deposition in the cell wall of *Bacillus subtilis*. *Journal of bacteriology*, 141 (2): 876-887].

[241] Zhang, S., Chen, X., Gu, C., Zhang, Y., Xu, J., Bian, Z., Yang, D. and Gu, N. (2009). The effect of iron oxide magnetic nanoparticles on smooth muscle cells. *Nanoscale Research Letters*, 4 (1): 70-77].

[242] Speranza, G., Gottardi, G., Pederzolli, C., Lunelli, L., Canteri, R., Pasquardini, L., Carli, E., Lui, A., Maniglio, D. and Brugnara, M. (2004). Role of chemical interactions in bacterial adhesion to polymer surfaces. *Biomaterials*, 25 (11): 2029-2037].

[243] Mohanraj, V. and Chen, Y. (2006). Nanoparticles-a review. *Tropical journal of pharmaceutical research*, 5 (1): 561-573].

[244] Sirelkhatim, A., Mahmud, S., Seeni, A., Kaus, N. H. M., Ann, L. C., Bakhori, S. K. M., Hasan, H. and Mohamad, D. (2015). Review on zinc oxide nanoparticles: antibacterial activity and toxicity mechanism. *Nano-micro letters*, 7 (3): 219-242].

[245] Qi, L., Xu, Z., Jiang, X., Hu, C. and Zou, X. (2004). Preparation and antibacterial activity of chitosan nanoparticles. *Carbohydrate research*, 339 (16): 2693-2700].

[246] Hu, X., Kandasamy Saravanakumar, T. J. and Wang, M. H. (2019). Mycosynthesis, characterization, anticancer and antibacterial activity of silver nanoparticles from endophytic fungus *Talaromyces purpureogenus*. *International Journal of Nanomedicine*, 14: 3427].

[247] Rani, R., Sharma, D., Chaturvedi, M. and Yadav, J. (2017). Green synthesis, characterization and antibacterial activity of silver nanoparticles of endophytic fungi *Aspergillus terreus*. *J. Nanomed. Nanotechnol*, 8 (4).

[248] Ahluwalia, V., Kumar, J., Sisodia, R., Shakil, N. A. and Walia, S. (2014). Green synthesis of silver nanoparticles by *Trichoderma harzianum* and their bio-efficacy evaluation against Staphylococcus aureus and Klebsiella pneumonia. *Industrial Crops and Products*, 55: 202-206].

[249] Bhat, M. A., Nayak, B. and Nanda, A. (2015). Evaluation of bactericidal activity of biologically synthesised silver nanoparticles from *Candida albicans* in combination with ciprofloxacin. *Materials Today: Proceedings*, 2 (9): 4395-4401].

[250] Guilger-Casagrande, M. and de Lima, R. (2019). Synthesis of silver nanoparticles mediated by fungi: A Review. *Frontiers in bioengineering and biotechnology*, 7.

[251] Jaidev, L. and Narasimha, G. (2010). Fungal mediated biosynthesis of silver nanoparticles, characterization and antimicrobial activity. *Colloids and surfaces B: Biointerfaces*, 81 (2): 430-433].

[252] Guilger-Casagrande, M. and de Lima, R. (2019). Fungal mediated synthesis of silver nanoparticles: A review. *Frontiers in Bioengineering and Biotechnology*, 7: 287].

[253] Ibrahim, H. M. and Hassan, M. S. (2016). Characterization and antimicrobial properties of cotton fabric loaded with green synthesized silver nanoparticles. *Carbohydrate polymers*, 151: 841-850].

[254] Singh, D., Rathod, V., Ninganagouda, S., Hiremath, J., Singh, A. K. and Mathew, J. (2014). Optimization and characterization of silver nanoparticle by endophytic fungi *Penicillium* sp. isolated from *Curcuma longa* (turmeric) and application studies against MDR *E. coli* and *S. aureus*. *Bioinorganic chemistry and applications*, 2014.

[255] Husseiny, S. M., Salah, T. A. and Anter, H. A. (2015). Biosynthesis of size controlled silver nanoparticles by *Fusarium oxysporum*, their antibacterial and antitumor activities. *Beni-Suef University Journal of Basic and Applied Sciences*, 4 (3): 225-231].

[256] Katas, H., Lim, C. S., Azlan, A. Y. H. N., Buang, F. and Busra, M. F. M. (2019). Antibacterial activity of biosynthesized gold nanoparticles using biomolecules from *Lignosus rhinocerotis* and chitosan. *Saudi Pharmaceutical Journal*, 27 (2): 283-292]

[257] El Domany, E. B., Essam, T. M., Ahmed, A. E. and Farghali, A. A. (2018). Biosynthesis physico-chemical optimization of gold nanoparticles as anti-cancer and synergetic antimicrobial activity using *Pleurotus ostreatus* fungus. *J. Appl. Pharm. Sci*, 2018: 8].

[258] Priyadarshini, E., Pradhan, N., Sukla, L. B. and Panda, P. K. (2014). Controlled synthesis of gold nanoparticles using *Aspergillus terreus* IF0 and its antibacterial potential against Gram negative pathogenic bacteria. *Journal of Nanotechnology*, 2014.

[259] Joshi, C. G., Danagoudar, A., Poyya, J., Kudva, A. K. and Dhananjaya, B. (2017). Biogenic synthesis of gold nanoparticles by marine endophytic fungus-*Cladosporium cladosporioides* isolated

from seaweed and evaluation of their antioxidant and antimicrobial properties. *Process Biochemistry*, 63: 137-144].

[260] Tripathi, R. M., Shrivastav, B. R. and Shrivastav, A. (2018). Antibacterial and catalytic activity of biogenic gold nanoparticles synthesised by *Trichoderma harzianum. IET nanobiotechnology*, 12 (4): 509-513].

[261] Balakumaran, M., Ramachandran, R., Balashanmugam, P., Mukeshkumar, D. and Kalaichelvan, P. (2016). Mycosynthesis of silver and gold nanoparticles: optimization, characterization and antimicrobial activity against human pathogens. *Microbiological research*, 182: 8-20].

[262] Soni, N. and Prakash, S. (2012). Synthesis of gold nanoparticles by the fungus *Aspergillus niger* and its efficacy against mosquito larvae. *Reports in Parasitology*, 2: 1-7].

[263] Abdeen, S., Isaac, R. R., Geo, S., Sornalekshmi, S., Rose, A. and Praseetha, P. (2013). Evaluation of Antimicrobial Activity of Biosynthesized Iron and Silver Nanoparticles Using the Fungi *Fusarium oxysporum* and *Actinomycetes* sp. on Human Pathogens. *Nano Biomedicine & Engineering*, 5 (1).

[264] Shankar, S. S., Rai, A., Ankamwar, B., Singh, A., Ahmad, A. and Sastry, M. (2004). Biological synthesis of triangular gold nanoprisms. *Nature materials*, 3 (7): 482-488].

[265] Bar, H., Bhui, D. K., Sahoo, G. P., Sarkar, P., De, S. P. and Misra, A. (2009). Green synthesis of silver nanoparticles using latex of *Jatropha curcas. Colloids and surfaces A: Physicochemical and engineering aspects*, 339 (1-3): 134-139].

[266] Parashar, U. K., Saxena, P. S. and Srivastava, A. (2009). Bioinspired synthesis of silver nanoparticles. *Digest Journal of Nanomaterials & Biostructures (DJNB)*, 4 (1).

[267] Philip, D. (2010). Green synthesis of gold and silver nanoparticles using *Hibiscus rosa sinensis. Physica E: Low-Dimensional Systems and Nanostructures*, 42 (5): 1417-1424].

[268] Govindaraju, K., Tamilselvan, S., Kiruthiga, V. and Singaravelu, G. (2010). Biogenic silver nanoparticles by Solanum torvum and their

promising antimicrobial activity. *Journal of Biopesticides*, 3 (1): 394].

[269] Nabikhan, A., Kandasamy, K., Raj, A. and Alikunhi, N. M. (2010). Synthesis of antimicrobial silver nanoparticles by callus and leaf extracts from saltmarsh plant, *Sesuvium portulacastrum* L *Colloids and surfaces B: Biointerfaces*, 79 (2): 488-493].

[270] Krishnaraj, C., Jagan, E., Rajasekar, S., Selvakumar, P., Kalaichelvan, P. and Mohan, N. (2010). Synthesis of silver nanoparticles using *Acalypha indica* leaf extracts and its antibacterial activity against water borne pathogens. *Colloids and Surfaces B: Biointerfaces*, 76 (1): 50-56].

[271] Dwivedi, A. D. and Gopal, K. (2010). Biosynthesis of silver and gold nanoparticles using *Chenopodium album* leaf extract. *Colloids and Surfaces A: Physicochemical and Engineering Aspects*, 369 (1-3): 27-33].

[272] Kouvaris, P., Delimitis, A., Zaspalis, V., Papadopoulos, D., Tsipas, S. A. and Michailidis, N. (2012). Green synthesis and characterization of silver nanoparticles produced using *Arbutus unedo* leaf extract. *Materials Letters*, 76: 18-20].

[273] Zahir, A. A., Bagavan, A., Kamaraj, C., Elango, G. and Rahuman, A. A. (2012). Efficacy of plant-mediated synthesized silver nanoparticles against *Sitophilus oryzae*. *Journal of Biopesticides*, 5: 95].

[274] Gnanajobitha, G., Paulkumar, K., Vanaja, M., Rajeshkumar, S., Malarkodi, C., Annadurai, G. and Kannan, C. (2013). Fruit-mediated synthesis of silver nanoparticles using *Vitis vinifera* and evaluation of their antimicrobial efficacy. *Journal of Nanostructure in Chemistry*, 3 (1): 67].

[275] Rodríguez-León, E., Iñiguez-Palomares, R., Navarro, R. E., Herrera-Urbina, R., Tánori, J., Iñiguez-Palomares, C. and Maldonado, A. (2013). Synthesis of silver nanoparticles using reducing agents obtained from natural sources (*Rumex hymenosepalus* extracts). *Nanoscale research letters*, 8 (1): 318].

[276] Biswas, K., Mohanta, Y. K., Kumar, V. B., Hashem, A., Abd_Allah, E. F., Mohanta, D. and Mohanta, T. K. (2020). Nutritional assessment study and role of green silver nanoparticles in shelf-life of coconut endosperm to develop as functional food. *Saudi Journal of Biological Sciences*.

[277] Geetha, R., Ashokkumar, T., Tamilselvan, S., Govindaraju, K., Sadiq, M. and Singaravelu, G. (2013). Green synthesis of gold nanoparticles and their anticancer activity. *Cancer nanotechnology*, 4 (4-5): 91-98].

[278] MubarakAli, D., Thajuddin, N., Jeganathan, K. and Gunasekaran, M. (2011). Plant extract mediated synthesis of silver and gold nanoparticles and its antibacterial activity against clinically isolated pathogens. *Colloids and Surfaces B: Biointerfaces*, 85 (2): 360-365].

[279] Folorunso, A., Akintelu, S., Oyebamiji, A. K., Ajayi, S., Abiola, B., Abdusalam, I. and Morakinyo, A. (2019). Biosynthesis, characterization and antimicrobial activity of gold nanoparticles from leaf extracts of *Annona muricata*. *Journal of Nanostructure in Chemistry*, 9 (2): 111-117].

[280] Wang, L., Xu, J., Yan, Y., Liu, H., Karunakaran, T. and Li, F. (2019). Green synthesis of gold nanoparticles from *Scutellaria barbata* and its anticancer activity in pancreatic cancer cell (PANC-1). *Artificial cells, nanomedicine, and biotechnology*, 47 (1): 1617-1627].

[281] Islam, N. U., Jalil, K., Shahid, M., Rauf, A., Muhammad, N., Khan, A., Shah, M. R. and Khan, M. A. (2019). Green synthesis and biological activities of gold nanoparticles functionalized with *Salix alba*. *Arabian Journal of Chemistry*, 12 (8): 2914-2925].

[282] Niraimathee, V., Subha, V., Ravindran, R. E. and Renganathan, S. (2016). Green synthesis of iron oxide nanoparticles from *Mimosa pudica* root extract. *International Journal of Environment and Sustainable Development*, 15 (3): 227-240].

[283] Latha, N. and Gowri, M. (2014). Biosynthesis and characterisation of Fe3O4 nanoparticles using Caricaya papaya leaves extract. *Int J Sci Res*, 3 (11): 1551-1556].

[284] Wang, Z., Fang, C. and Mallavarapu, M. (2015). Characterization of iron–polyphenol complex nanoparticles synthesized by Sage (*Salvia officinalis*) leaves. *Environmental Technology & Innovation*, 4: 92-97].

[285] Machado, S., Pacheco, J., Nouws, H., Albergaria, J. T. and Delerue-Matos, C. (2015). Characterization of green zero-valent iron nanoparticles produced with tree leaf extracts. *Science of the total environment*, 533: 76-81].

[286] Jassal, V., Shanker, U. and Gahlot, S. (2016). Green synthesis of some iron oxide nanoparticles and their interaction with 2-amino, 3-amino and 4-aminopyridines. *Materials Today: Proceedings*, 3 (6): 1874-1882].

[287] Cao, D., Jin, X., Gan, L., Wang, T. and Chen, Z. (2016). Removal of phosphate using iron oxide nanoparticles synthesized by eucalyptus leaf extract in the presence of CTAB surfactant. *Chemosphere*, 159: 23-31].

[288] Chen, G., Roy, I., Yang, C. and Prasad, P. N. (2016). Nanochemistry and nanomedicine for nanoparticle-based diagnostics and therapy. *Chemical reviews*, 116 (5): 2826-2885].

[289] Muthukumar, H. and Matheswaran, M. (2015). Amaranthus spinosus leaf extract mediated FeO nanoparticles: physicochemical traits, photocatalytic and antioxidant activity. *ACS Sustainable Chemistry & Engineering*, 3 (12): 3149-3156].

[290] Wang, T., Jin, X., Chen, Z., Megharaj, M. and Naidu, R. (2014). Green synthesis of Fe nanoparticles using eucalyptus leaf extracts for treatment of eutrophic wastewater. *Science of the total environment*, 466: 210-213].

[291] Kozma, G., Rónavári, A., Kónya, Z. and Kukovecz, A. (2016). Environmentally benign synthesis methods of zero-valent iron nanoparticles. *ACS Sustainable Chemistry & Engineering*, 4 (1): 291-297].

[292] Toli, A., Chalastara, K., Mystrioti, C., Xenidis, A. and Papassiopi, N. (2016). Incorporation of zero valent iron nanoparticles in the

matrix of cationic resin beads for the remediation of Cr (VI) contaminated waters. *Environmental Pollution*, 214: 419-429].

[293] Khanna, P., Gaikwad, S., Adhyapak, P., Singh, N. and Marimuthu, R. (2007). Synthesis and characterization of copper nanoparticles. *Materials Letters*, 61 (25): 4711-4714].

[294] Subhankari, I. and Nayak, P. (2013). Synthesis of copper nanoparticles using *Syzygium aromaticum* (Cloves) aqueous extract by using green chemistry. *World J Nano Sci Technol*, 2 (1): 14-17].

[295] Sadanand, V., Rajini, N., Rajulu, A. V. and Satyanarayana, B. (2016). Preparation of cellulose composites with in situ generated copper nanoparticles using leaf extract and their properties. *Carbohydrate polymers*, 150: 32-39].

[296] Din, M. I., Arshad, F., Hussain, Z. and Mukhtar, M. (2017). Green adeptness in the synthesis and stabilization of copper nanoparticles: catalytic, antibacterial, cytotoxicity, and antioxidant activities. *Nanoscale research letters*, 12 (1): 638].

[297] Hussain, I., Singh, N., Singh, A., Singh, H. and Singh, S. (2016). Green synthesis of nanoparticles and its potential application. *Biotechnology letters*, 38 (4): 545-560].

[298] Shende, S., Ingle, A. P., Gade, A. and Rai, M. (2015). Green synthesis of copper nanoparticles by *Citrus medica* Linn. (Idilimbu) juice and its antimicrobial activity. *World Journal of Microbiology and Biotechnology*, 31 (6): 865-873].

[299] Nasrollahzadeh, M., Sajadi, S. M. and Khalaj, M. (2014). Green synthesis of copper nanoparticles using aqueous extract of the leaves of *Euphorbia esula* L and their catalytic activity for ligand-free Ullmann-coupling reaction and reduction of 4-nitrophenol. *RSC Advances*, 4 (88): 47313-47318].

[300] Awwad, A. M. and Albiss, B. (2015). Biosynthesis of colloidal copper hydroxide nanowires using pistachio leaf extract. *Advanced Materials Letters*, 6 (1): 51-54].

[301] Awwad, A., Albiss, B. and Salem, N. (2015). Antibacterial activity of synthesized copper oxide nanoparticles using *Malva sylvestris* leaf extract. *SMU Med J*, 2 (1): 91-101].

[302] de Aragao, A. P., de Oliveira, T. M., Quelemes, P. V., Perfeito, M. L. G., Araujo, M. C., Santiago, J. d. A. S., Cardoso, V. S., Quaresma, P., de Almeida, J. R. d. S. and da Silva, D. A. (2019). Green synthesis of silver nanoparticles using the seaweed *Gracilaria birdiae* and their antibacterial activity. *Arabian Journal of Chemistry*, 12 (8): 4182-4188].

[303] Senthilkumar, R., Reddy Prasad, D., Govindarajan, L., Saravanakumar, K. and Naveen Prasad, B. (2020). Synthesis of green marine algal-based biochar for remediation of arsenic (V) from contaminated waters in batch and column mode of operation. *International Journal of Phytoremediation*, 22 (3): 279-286].

[304] Balaji, D., Basavaraja, S., Deshpande, R., Mahesh, D. B., Prabhakar, B. and Venkataraman, A. (2009). Extracellular biosynthesis of functionalized silver nanoparticles by strains of *Cladosporium cladosporioides* fungus. *Colloids and surfaces B: biointerfaces*, 68 (1): 88-92].

[305] Sanghi, R. and Verma, P. (2009). Biomimetic synthesis and characterisation of protein capped silver nanoparticles. *Bioresource technology*, 100 (1): 501-504].

[306] Senapati, S., Mandal, D., Ahmad, A., Khan, M. I., Sastry, M. and Kumar, R. (2004). Fungus mediated synthesis of silver nanoparticles: a novel biological approach. *Indian Journal of Physics*, 78: 101-105].

[307] Chowdhury, S., Basu, A. and Kundu, S. (2014). Green synthesis of protein capped silver nanoparticles from phytopathogenic fungus *Macrophomina phaseolina* (Tassi) Goid with antimicrobial properties against multidrug-resistant bacteria. *Nanoscale research letters*, 9 (1): 1-11].

[308] Shaligram, N. S., Bule, M., Bhambure, R., Singhal, R. S., Singh, S. K., Szakacs, G. and Pandey, A. (2009). Biosynthesis of silver nanoparticles using aqueous extract from the compactin producing fungal strain. *Process biochemistry*, 44 (8): 939-943].

[309] Kathiresan, K., Manivannan, S., Nabeel, M. and Dhivya, B. (2009). Studies on silver nanoparticles synthesized by a marine fungus,

Penicillium fellutanum isolated from coastal mangrove sediment. *Colloids and surfaces B: Biointerfaces*, 71 (1): 133-137].

[310] Maliszewska, I., Juraszek, A. and Bielska, K. (2014). Green synthesis and characterization of silver nanoparticles using ascomycota fungi *Penicillium nalgiovense* AJ12. *Journal of Cluster Science*, 25 (4): 989-1004].

[311] Vigneshwaran, N., Kathe, A. A., Varadarajan, P. V., Nachane, R. P. and Balasubramanya, R. H. (2007). Silver– protein (core– shell) nanoparticle production using spent mushroom substrate. *Langmuir*, 23 (13): 7113-7117].

[312] Birla, S., Tiwari, V., Gade, A., Ingle, A., Yadav, A. and Rai, M. (2009). Fabrication of silver nanoparticles by Phoma glomerata and its combined effect against *Escherichia coli*, *Pseudomonas aeruginosa* and *Staphylococcus aureus*. *Letters in Applied Microbiology*, 48 (2): 173-179].

[313] Fayaz, M., Tiwary, C., Kalaichelvan, P. and Venkatesan, R. (2010). Blue orange light emission from biogenic synthesized silver nanoparticles using *Trichoderma viride*. *Colloids and Surfaces B: Biointerfaces*, 75 (1): 175-178].

[314] Binupriya, A., Sathishkumar, M., Vijayaraghavan, K. and Yun, S. I. (2010). Bioreduction of trivalent aurum to nano-crystalline gold particles by active and inactive cells and cell-free extract of *Aspergillus oryzae* var. viridis. *Journal of Hazardous Materials*, 177 (1-3): 539-545].

[315] Tidke, P. R., Gupta, I., Gade, A. K. and Rai, M. (2014). Fungus-mediated synthesis of gold nanoparticles and standardization of parameters for its biosynthesis. *IEEE transactions on nanobioscience*, 13 (4): 397-402].

[316] Honary, S., Gharaei-Fathabad, E., Barabadi, H. and Naghibi, F. (2013). Fungus-mediated synthesis of gold nanoparticles: a novel biological approach to nanoparticle synthesis. *Journal of nanoscience and nanotechnology*, 13 (2): 1427-1430].

[317] Mishra, A., Tripathy, S. K. and Yun, S. I. (2012). Fungus mediated synthesis of gold nanoparticles and their conjugation with genomic

DNA isolated from *Escherichia coli* and *Staphylococcus aureus*. *Process Biochemistry*, 47 (5): 701-711].

[318] Soni, N. and Prakash, S. (2012). Efficacy of fungus mediated silver and gold nanoparticles against *Aedes aegypti* larvae. *Parasitology research*, 110 (1): 175-184].

[319] Chauhan, A., Zubair, S., Tufail, S., Sherwani, A., Sajid, M., Raman, S. C., Azam, A. and Owais, M. (2011). Fungus-mediated biological synthesis of gold nanoparticles: potential in detection of liver cancer. *International journal of nanomedicine*, 6: 2305].

[320] Verma, V. C., Singh, S. K., Solanki, R. and Prakash, S. (2011). Biofabrication of anisotropic gold nanotriangles using extract of endophytic *Aspergillus clavatus* as a dual functional reductant and stabilizer. *Nanoscale Res Lett*, 6 (1): 1-7].

[321] Ammar, H. A., Rabie, G. H. and Mohamed, E. (2019). Novel fabrication of gelatin-encapsulated copper nanoparticles using *Aspergillus versicolor* and their application in controlling of rotting plant pathogens. *Bioprocess and biosystems engineering*, 42 (12): 1947-1961].

[322] Pavani, K. and Kumar, N. S. (2013). Adsorption of iron and synthesis of iron nanoparticles by *Aspergillus species* kvp 12. *Am J Nanomater*, 1 (2): 24-26].

[323] Gade, A., Ingle, A., Whiteley, C. and Rai, M. (2010). Mycogenic metal nanoparticles: progress and applications. *Biotechnology letters*, 32 (5): 593-600].

[324] Saif, S., Tahir, A. and Chen, Y. (2016). Green synthesis of iron nanoparticles and their environmental applications and implications. *Nanomaterials*, 6 (11): 209].

[325] Lakshmanan, G., Sathiyaseelan, A., Kalaichelvan, P. and Murugesan, K. (2018). Plant-mediated synthesis of silver nanoparticles using fruit extract of *Cleome viscosa* L.: Assessment of their antibacterial and anticancer activity. *Karbala International Journal of Modern Science*, 4 (1): 61-68].

[326] Keihan, A. H., Veisi, H. and Veasi, H. (2017). Green synthesis and characterization of spherical copper nanoparticles as organometallic

antibacterial agent. *Applied Organometallic Chemistry*, 31 (7): e3642].

[327] Parihar, G. and Balekar, N. (2016). *Calotropis procera*: A phytochemical and pharmacological review. *Thai Journal of Pharmaceutical Sciences*, 40 (3).

[328] Manikprabhu, D., Cheng, J., Chen, W., Sunkara, A. K., Mane, S. B., Kumar, R., Hozzein, W. N., Duan, Y. Q. and Li, W. J. (2016). Sunlight mediated synthesis of silver nanoparticles by a novel actinobacterium (*Sinomonas mesophila* MPKL 26) and its antimicrobial activity against multi drug resistant Staphylococcus aureus. *Journal of Photochemistry and Photobiology B: Biology*, 158: 202-205].

[329] Hindi, N. K. K. and Chabuck, Z. A. G. (2013). Antimicrobial activity of different aqueous lemon extracts. *Journal of Applied Pharmaceutical Science*, 3 (6): 74].

[330] Lee, H. J., Song, J. Y. and Kim, B. S. (2013). Biological synthesis of copper nanoparticles using *Magnolia kobus* leaf extract and their antibacterial activity.*Journal of Chemical Technology & Biotechnology*, 88 (11): 1971-1977].

[331] Gopinath, M., Subbaiya, R., Selvam, M. M. and Suresh, D. (2014). Synthesis of copper nanoparticles from *Nerium oleander* leaf aqueous extract and its antibacterial activity. *Int J Curr Microbiol App Sci*, 3 (9): 814-818].

[332] Vitta, Y., Figueroa, M., Calderon, M. and Ciangherotti, C. (2020). Synthesis of iron nanoparticles from aqueous extract of *Eucalyptus robusta* Sm and evaluation of antioxidant and antimicrobial activity. *Materials Science for Energy Technologies*, 3: 97-103].

[333] Saranya, S., Vijayarani, K. and Pavithra, S. (2017). Green synthesis of iron nanoparticles using aqueous extract of *Musa ornata* flower sheath against pathogenic bacteria. *Indian Journal of Pharmaceutical Sciences*, 79 (5): 688-694].

[334] Maghsoudy, N., Azar, P. A., Tehrani, M. S., Husain, S. W. and Larijani, K. (2019). Biosynthesis of Ag and Fe nanoparticles using *Erodium cicutarium*; study, optimization, and modeling of the

antibacterial properties using response surface methodology. *Journal of Nanostructure in Chemistry*, 9 (3): 203-216].

[335] Jagathesan, G. and Rajiv, P. (2018). Biosynthesis and characterization of iron oxide nanoparticles using *Eichhornia crassipes* leaf extract and assessing their antibacterial activity. *Biocatalysis and agricultural biotechnology*, 13: 90-94].

[336] Kathiraven, T., Sundaramanickam, A., Shanmugam, N. and Balasubramanian, T. (2015). Green synthesis of silver nanoparticles using marine algae *Caulerpa racemosa* and their antibacterial activity against some human pathogens. *Applied Nanoscience*, 5 (4): 499-504].

[337] Abdel-Raouf, N., Al-Enazi, N. M. and Ibraheem, I. B. (2017). Green biosynthesis of gold nanoparticles using *Galaxaura elongata* and characterization of their antibacterial activity. *Arabian Journal of Chemistry*, 10: S3029-S3039].

[338] Gade, A., Bonde, P., Ingle, A., Marcato, P., Duran, N. and Rai, M. (2008). Exploitation of *Aspergillus niger* for synthesis of silver nanoparticles. *Journal of Biobased Materials and Bioenergy*, 2 (3): 243-247].

[339] Ingle, A., Gade, A., Pierrat, S., Sonnichsen, C. and Rai, M. (2008). Mycosynthesis of silver nanoparticles using the fungus *Fusarium acuminatum* and its activity against some human pathogenic bacteria. *Current Nanoscience*, 4 (2): 141-144].

[340] Elumalai, S., Devika, R., Arumugam, P. and Kasinathan, K. (2012). Extracellular biosynthesis of silver nanoparticles using the fungus pleurotus ostreatus and their antibacterial activity. *Journal of NanoScience, NanoEngineering & Applications*, 2 (2).

[341] Ray, S., Sarkar, S. and Kundu, S. (2011). Extracellular biosynthesis of silver nanoparticles using the mycorrhizal mushroom *Tricholoma crassum* (Berk.) Sacc: its antimicrobial activity against pathogenic bacteria and fungus, including multidrug resistant plant and human bacteria. *Dig J Nanomater Biostruct*, 6: 1289-1299].

[342] Jalal, M., Ansari, M. A., Alzohairy, M. A., Ali, S. G., Khan, H. M., Almatroudi, A. and Raees, K. (2018). Biosynthesis of silver

nanoparticles from *oropharyngeal candida glabrata* isolates and their antimicrobial activity against clinical strains of bacteria and fungi. *Nanomaterials*, 8 (8): 586].

[343] Ahmed, A. A., Hamzah, H. and Maaroof, M. (2018). Analyzing formation of silver nanoparticles from the filamentous fungus *Fusarium oxysporum* and their antimicrobial activity. *Turkish Journal of Biology*, 42 (1): 54-62].

[344] Singh, P. S. and Vidyasagar, G. (2018). Biosynthesis of antibacterial silver nano-particles from *Aspergillus terreus*. *World News of Natural Sciences*, 16: 117-124].

[345] Mohanta, Y. K., Nayak, D., Biswas, K., Singdevsachan, S. K., Abd_Allah, E. F., Hashem, A., Alqarawi, A. A., Yadav, D. and Mohanta, T. K. (2018). Silver nanoparticles synthesized using wild mushroom show potential antimicrobial activities against food borne pathogens. *Molecules*, 23 (3): 655].

[346] Ma, L., Su, W., Liu, J. X., Zeng, X. X., Huang, Z., Li, W., Liu, Z. C. and Tang, J. X. (2017). Optimization for extracellular biosynthesis of silver nanoparticles by *Penicillium aculeatum* Su1 and their antimicrobial activity and cytotoxic effect compared with silver ions. *Materials Science and Engineering: C*, 77: 963-971].

[347] Soliman, H., Elsayed, A. and Dyaa, A. (2018). Antimicrobial activity of silver nanoparticles biosynthesised by *Rhodotorula* sp. strain ATL72. *Egyptian Journal of Basic and Applied Sciences*, 5 (3): 228-233].

In: Recent Developments in *Enterobacter* … ISBN: 978-1-53618-615-4
Editor: Grégoire Roux

Chapter 3

Latest Treatment Strategies against Multi-Drug Resistant Enterobacteriaceae Infections

Hafiz Muhammad Tahir, Muhammad Summer and Shaukat Ali
Department of Zoology,
Government College University Lahore, Pakistan

Abstract

During the last few decades, the antimicrobial resistance due to overuse and misuse of antibiotics has become significant and springing as a grave risk to global health concerns. The rising rate of multiple drug resistance (MDR) is reducing the reliability of first-line traditional available pharmaceutics like cephalosporins, penicillins and fluoroquinolones for the treatment of community and hospital-acquired (nosocomial) infections caused by the pathogens of family Enterobacteriaceae. High rates of morbidity and mortality alongwith the prolonged hospitalization are common fates in such bacterial contagiousness. To overcome the complexities regarding the antibiotic therapies, MDR and to combat the pathogens, there is a dire need to develop some novel clinical strategies.

In this perspective, the world scientific community suggests the combined use of multiple drugs that target peptidoglycan synthesis as compare to monotherapy, which shows promising results in combating the infections caused by Enterobacteriaceae. Moreover, various antibiotics like polymyxins, fosfomycin and tigecycline (semisynthetic glycylcycline) have also provided the efficacious therapeutic alternatives to clinicians for critically ill hosts. Polymyxin B and colistin are remerged as a last resort therapeutics for the treatment of nosocomial infections caused by superbugs (pathogens which are almost resistant to every existing antibiotic). Furthermore, cationic antimicrobial peptides via autolysis and enzyme synthesis blockage, peptidomimetics and foldamers (non-toxic small peptides with amphiphilic nature) via membrane lysis, siderophores and sideromycins via depletion of iron needed for the growth of bacteria, efflux pump inhibitors via high intercellular drug persistency and therapeutic antibodies via phagocytosis to neutralize the different strains like *Enterococcus, Escherichia coli* and *Klabsiella pneumonia*(Enterobacteriaceae).

In addition to the on-deck antibiotics, clinicians and medicinal experts are intended to revive the use of old and abandoned drugs like temocillin and fosfomycin with modernistic and different combinations and strategies to treat the infections of Enterobacteriaceae. Moreover, bacteriophages (natural predators of bacteria) are again in use for the treatment of different MDR pathogenic organisms like *Pseudeomonas aeruginosa* and *Staphylococcus aureus*. As the MDR arose due to the presence of lactamases in the pathogens, the global health sector is in pipeline to design the β-lactam/β-lactamase inhibitors with significant and promising *in vitro* activity to combat the MDR pathogens. Further deep analysis with the clinical approach is needed along with the prolific and practical strategies to formulate the suitable drug candidates to cope up with the battle against the emerging MDR.

1. INTRODUCTION

Enterobacteriaceae is a large family of Gram-negative bacteria. This family was proposed by Rahn in 1936 which includes over 30 genera and 100 species. Some human pathogenic bacteria like *Klebsiella pneumoniae*, *Escherichia coli*, *Proteus mirabilis*, *Salmonella enterica*, *Raoultella planticola* and *Citrobacter freundii* belong to this family (Potter et al.,

2016). Among these, *K. pneumonia* and *E. coli* are evaluated are considered as the root cause of community-based and nosocomial infections globally. These infections have become a grave clinical and economic challenge (Peterson, 2006; Sanyasi et al., 2016; Wright et al., 2017). Although, the dawn of an antibiotic era in the mid of 20th century revolutionized the medical care strategies (Wright et al., 2017) due to unregulated and overuse of these frontline antibiotics, efficacy has been constantly decreasing and different bacterial strains have developed resistance (Levy & Marshall, 2004; Arvizo et al., 2012; Xu et al., 2014). Multiple drug resistance (MDR) has emerged as a potential hurdle in the treatment of the infections caused by Enterobacteriaceae (Franci et al., 2015). The phenomenon of MDR has endangered the effectiveness of different drugs like cephalosporins, penicillins and fluoroquinolones, etc. which had saved millions of precious lives all over the world (Gould &Bal, 2013; Golkar et al., 2014).

2. Causes of MDR

Multiple key factors like overuse, inappropriate prescribing, availability and intensive agricultural use are involved in the development of multiple drug resistance.

2.1. Overuse

Regarding the overuse of antibiotics, Sir Alexander Fleming had warned the world as "Public will demand [the drug] and then will begin an era of abuses (Spellberg & Gilbert, 2014; Bartlett et al., 2013). Within a short time, unregulated and overuse of synthetic drugs has evolved the resistance (Read & Woods, 2014). An epidemiological study globally revealed the positive correlation between the consumption of the antibiotics and the developing of resistant bacterial strains (Gross, 2012). Moreover, in many developing and underdeveloped countries, the majority of antibiotics are accessible to the public without a prescription which encourages overuse (Michael et al., 2014). Genes can be passed intraspecifically or can be acquired interspecifically through plasmids in bacteria. This phenomenon of horizontal gene transfer, transports the resistant genes to the next generations of different bacteria (Read & Woods, 2014).

2.2. Unprofessional Prescription

Inappropriate prescription is another major factor behind the emergence of resistance among the variety of human and non-human pathogens (Marc et al., 2016). After significant research, the world scientific community has concluded that antibiotic selection and duration of treatment are30 to 50% incorrect throughout the world (Luyt et al., 2014). Furthermore, non-essential, incorrect and insignificant drugs are being prescribed in the intensive care units (Luyt et al., 2014). This scenario leads to the arguable clinical outcomes and sometimes severe side effects in the antibiotic-treated patients (Lushniak, 2014). In some other cases, antibiotic resistance could also promote by mutagenesis, variation in gene expressions and HGT (Viswanathan, 2014).

2.3. Substantial Use in Agriculture Sector

Antibiotics are being used extensively throughout the world for acquiring the optimal growth of livestock as well as for infection remedy, higher product yields with prime quality (Bartlett et al., 2014; Spellberg & Gilbert, 2014; Gross, 2013; Michael et al., 2014). Antibiotics administered to the animals passed onto humans through the consumption of food and milk products. Similarly, resistant strains also transferred from animals to consumers through meat consumption that causes serious infections and detrimental health consequences (Golkar et al., 2014; Rossolini et al., 2014). Moreover, synthetic antibacterial products used for the hygienic purpose also pileup the resistance which ultimately leads to the hijacking of the immune system and morbidity (Piddock, 2012; Golkar et al., 2014).

2.4. Scarcity of Novel Antibiotics

The pace of new drug fabrication by the pharma industry is slowed down due to financial and regulatory hurdles (antibiotics approval, etc.) which had been a successful approach to combat the resistant pathogens (Bartlett et al., 2013; Bbosa et al., 2014). New research on the development of new antibiotics is also very slow due to the unavailability of funds to the institutions (Piddock, 2012; Ventola, 2015).

3. General Concept of Drug Resistance

Resistance against the various types of antibiotics could develop either by expression of chromosomal genes or transferred from one generation to next through plasmids (Kraiskos et al., 2019). There are several modes of resistance to the β-lactams (antibiotics that harbor a distinct lactam ring in their molecular composition) like modification of the target site, β-lactamases (hydrolytic enzymes) and efflux pumps (Ambler, 1980; Bush & Jacoby, 2010; Docquier & Mangani, 2018). During the development of resistance, the bacterial β-lactamases break the β-lactam ring through hydrolyzing the amide linkage (Bush & Jacoby, 2010; Delgado-Valverde, 2013; Docquier & Mangani, 2018; Kraiskos et al., 2019). B-lactamases have four classes (A, B, C and D) depending upon the variability of molecular structure. Enzymes that hydrolyze the serine belong to class A, C and D while zinc mettalo β-lactamases constitute the class B (Queenan et al., 2007; Wright et al., 2017). Clinically significant β-lactamases included the (ESBL) extended-spectrum β-lactamases (TEM, SHV, CTX-M) carbapenemases and AmpC enzymes (Bush, 2018; Kraiskos et al., 2019).

4. Treatment Options for MDR Enterobacteriaceae Infections

Although multiple drug resistance has questioned the efficacy of the frontline pharmaceutics but still there are considerable remedial options for managing and combating the serious infections caused by the Enterobacteriaceae. In this chapter we discuss the available approaches for the treatment which are as follows:

Conventional Antibiotics

- β-lactam antibiotics
- β-lactamase inhibitors
- Tetracyclines
- Aminoglycosides
- Reuse of abandoned drugs

Non-Conventional Approaches

- Non-conventional antibiotics
- Phage therapy
- Combination therapy
- Nanoparticles (A modern way to combat MDR pathogens)

4.1. β-Lactam Antibiotics

These antibiotics have the bactericidal potential as they inhibit the growth of bacterial cell walls (Xu et al., 2014). The interruption of linkage between the glycosides component of the wall is the mode of action of β-lactam antibiotics (Donowitz & Mandell, 1988; Lahiri et al., 2016; Durand et al., 2017). Significant results have been reported against the Gram-negative bacterial strains like members of Enterobacteriaceae (Papp & Endimiani, 2011). One of the renowned members of this class of antibiotics are given below:

4.1.1. Carbapenems

Carbapenems are scrutinized as a suitable candidate for having efficient remedial outcomes against bacteria which produce the ESBL after interaction with the active ingredients of the drugs (Bassetti et al., 2007; Lopez-Cerero et al., 2010). A recent study has revealed the less mortality of patients treated with Carbapenems as compared to the other commonly used antibiotics like fluoroquinolones, cephalosporins and aminoglycosides (Vardakas et al., 2012). Imipenem and ertapenem are regarded as the appropriate members of Carbapenems for treatment of the infections acquired from the Gram-negative bacteria like *Acinetobacter baumanni* (Collins et al., 2012; Nicolau et al., 2012).

4.1.2. Cephalosporins

These drugs are categorized into 3rd and 4th generation cephalosporins which are being used against various types of pathogens (Delgado-Valverde et al., 2013). However, 3rd generation cephalosporins are less efficacious against severe infections (Choi et al., 2008; Harris & Ferguson, 2012). Although the 4th generation cephalosporins like cefepime and cefpirome are proved to be more effective against the different types of ESBL producing pathogens (Navarro et al., 2010).

4.2. β-Lactamase Inhibitors

Several types of β-lactamase inhibitors are under process and in clinical trials for the treatment of pathogenic illness (Xu et al., 2014). These compounds harbor the inhibitory potential against various types of bacterial enzymes and assist in the remitting the activity of other antibiotics like carbapenems and penicillins (Bebrone et al., 2010). Different members of β-lactamase inhibitors like 6-methylidene-penem and oxapenams which shows significant and broad-spectrum activity against class A, C and D while positive results are observed in different Gram-negative bacteria against piperacillin when used in combination (Ruzin et al., 2010). Avibactam is another non-beta lactam molecule which resists the formation of beta-lactamase of class A and C by formation of carbonyl linkage (Ehmann et al., 2012; Aktas et al., 2012; Louie et al., 2012). The antibacterial potential of avibactam against *E.coli* and *K. pneumonia* could be enhanced many folds by formulating its combination with different potent cephalosporins (Lagace et al., 2011; Livermore et al., 2012).

4.3. Tetracyclines

Eravacycline (tetracycline) is a novel and synthetic flourocycline molecule which can neutralize the bacteria by blocking the protein synthesis through binding with their ribosomes (Xiao et al., 2012; Grossman et al., 2012; Zhanel et al., 2016). This antibiotic is very effective against the various pathogens except for *P. aeruginosa* including MDREnterobacteriaceae and carbapenem, cephalosporins and fluoroquinolones resistant bacteria (Bassetti & Righi, 2014; Abdullah et al., 2015; Zhang et al., 2016; Livermore et al., 2016). The activity of this drug is remained unaffected by common resistant modes of the tetracyclines like tetracycline based efflux pumps and ribosomal protecting protein molecules (Sutcliffe et al., 2013; Bassetti & Righi, 2014; Xu et al., 2014). Phase II trials for intravenous eravacycline drug formation are in pipeline which could be possibly more efficacious than the present counterpart (Xu et al., 2014).

4.4. Aminoglycosides

Aminoglycosides are in use for the treatment of various health-related issues. Among aminoglycosides, plazomicin has also the ability to kill the Gram-negative bacteria (by inhibiting their protein synthesis through binding with one of the subunit (30S) of the ribosome (Wright et al., 2017). This drug has been formulized to overcome the problem of bacterial resistance against conventional glycosides as the majority of the pathogens are housed to certain aminoglycoside mutating enzymes (Armstrong et al., 2010; Aggen et al., 2010; Zhanel et al., 2012). Efficacy of this aminoglycoside could enhance many folds against carbapenem-resistant strains of Enterobacteriaceae when administered with a combination of gentamicin and tobramycin (Livermore et al., 2011; Galani et al., 2012; Wright et al., 2017).

4.5. Reuse of Abandoned Drugs

Including polymixins (polymixin B and E) which were widely used for combating the Gram-negative bacteria including MDR, Enterobacteriaceae have abandoned since 1980s. Now the world clinician's community is willing to revive their use alongwith some other old antibiotics like fosfomycin, temocillin and phage therapy (Landman et al., 2008; Kofteridis et al., 2010; Velkov et al., 2010; Delgado-valvaede et al., 2013). Temocillin is a β-lactam, derivative of ticarcillin, having a significant potential to kill the MDR (ESBL producing) pathogens except for the Gram-positive bacteria (Roux et al., 2012; Skurnik et al., 2012).

Fosfomycin, a unique antibacterial agent, is a derivative of phosphonic acid which was identified and extracted from the *Streptomyces* cultures (Michalopoulos et al., 2011; Karageorgopoulos et al., 2012). This antibiotic shows promising results against carbapenemase-producing Enterobacteriaceae including *K. pneumonia* in clinical trials (Michalopoulos et al., 2010). Efficacy of this drug could be enhanced by using it with other drugs in a combination against human pathogens (mostly MDR Gram-negative bacteria) causing pneumonia, meningitis and pyelonephritis as well as nosocomial infections (Falagas et al., 2011; Xu et al., 2014). Fosfomycin blocks the first step of peptidoglycan synthesis by irreversibly blocking the catalyzer enolpyruvyle transferase ultimately leading to the pores in the bacterial cell wall (Wittebole et al., 2014; Venugopal et al., 2012).

Tigecycline has the therapeutic benefits in serious and polymicrobial infections caused due to certain Gram-negative including

Enterobacteriaceae and Gram-positive bacterial strains (Milatovic et al., 2003; Noskin, 2005; Zhanel et al., 2009; Delgado-valverde et al., 2013). Tigecycline bind to the 30S subunit of the ribosome, hence block the aminoacyl entry into ribosomal A site and interrupt the process of translation in bacteria (Jones et al., 2007).

4.6. Non-Conventional Approaches

4.6.1. Natural Antibacterial Peptides (NAPs)

Natural cationic peptides that are considered as natural antibiotics have emerged as potential antimicrobial agents against bacteria, fungi and various types of viruses (Hancocke et al., 1999; Jenssen et al., 2006; Xu et al., 2014). Almost 1000 of these peptides have been extracted from natural sources like animals and plants and are already being used in the medical and food sectors (Hale et al., 2007). These defense peptides are comprised of 12-50 amino acids with hydrophobic nature and have the significant antibacterial potential for s variety of Gram-negative bacteria (Powers & Hancock, 2003; Hadley & Hancock, 2010; Findlay et al., 2010; Baltzer & Brown, 2011; Xu et al., 2014). Cationic peptides are preferred over conventional antibiotics as they have rapid disruptive action on bacterial membranes and disrupt cellular integrity (Hale, 2007; Xu et al., 2014). Omiganan and pexiganan are novel peptides have been in preclinical trials for the treatment of different microbe related illnesses and infections (Melo et al., 2006; Lipsky et al., 2008; Rubinchik et al., 2009; Yin et al., 2012; Lin et al., 2015). Furthermore, large scale synthesis of these peptides is feasible through chemical and recombinant approaches in a cost-effective manner (Bommarius et al., 2010).

4.6.2. Antimicrobial Peptidomimetics (AMPs)

To overcome the resistive demerits of the cationic AMPs, researchers have developed small, non-toxic peptide mimicking compounds and foldamers with antibacterial activity (Tew et al., 2010; Hu et al., 2011; Giuliani & Rinaldi, 2011; Molchanova et al., 2017). These molecules are being fabricated from existing AMPs and through sequence modifications and de novo approach (Giuliani et al., 2008; Fjell et al., 2011; Scorciapino & Andrea, 2012). Recently, peptidomimetic foldamer brilacidin has been designed which can bind to outer membranes of bacteria and showed bactericidal potential for MDR Gram-negative strains including *E. coli* and *K. pneumonia* (Thaker et al., 2011; Mensa et al., 2011; Scorciapino & Andrea, 2012).

A novel group of arylamide foldamers has been reported named α-AA peptides like PMX-10072 are potent bactericidal agents for family Enterobacteriaceae, certain Gram-positive strains and fungi (Yashima et al., 2016; Gopalakrishnan et al., 2016). These peptidomimetics are derivatives of the N-acetyl N-aminoethyl amino acids which are easy to manufacture and cost-effective (Padhee et al., 2011).

4.6.3. Iron Binding Glycoproteins

Iron is an essential component for bacterial life due to its structural as well as functional role like maintaining the homeostasis (Xu et al., 2014). In living organisms like humans, the majority of the iron is inbound form like hemoglobin, transferrin and lactoferrin which keep the bacteria at bay from causing infections due to unavailability of free iron for pathogens (Saha et al., 2016; Ito et al., 2016; Ito et al., 2018). To overcome the shortage of iron, bacteria release siderophores for iron fetching from the available sources. Talactoferrin (iron-binding glycoprotein of humans) has a wide range of abilities including maintaining the iron balance, protection against microbial infection and anti-inflammatory role (Kelly et al., 2010). Its antibacterial activity is a broad-spectrum, could inhibit the growth of bacteria, viruses and protozoa (Xu et al., 2014; Peters et al., 2020). These compounds either directly make the free iron unavailable in living systems for bacteria or direct cell disruption through surface attachment (Venkatesh & Rong, 2008; Liu et al., 2019).

Different other AMPs including EA-230 (hCG derivative), Hlf1-11 (lactoferrin derivative), PXLO1 (lactoferrin analogue), surotomycin (cyclic lipopeptidederivative), POL7080 (protein analogue) and many more are in phase III of clinical trials and could be used against MDR Enterobacteriaceae (Koo & Seo, 2019).

4.6.4. Efflux Pump Inhibitors (EPIs)

Word scientists have been working for20 years for synthesizing the potentially active EPIs for resisting the phenomenon of MDR, ultimately succeeded in the fabrication of reasonable drugs including reserpine and verapamil as EPIs (Aeschilmann et al., 1999; Nikaido, 2011; Lamut et al., 2019). Recently,efficacious and potent EPIs like PaβN, pyridopyrimidinone D-13-9001, pyranopyridine MBX2319 and different indole derivatives are identified and reported (Mahmood et al., 2016; Santajit & Indrawattana, 2016; Tacconelli et al., 2018). EPIs work by acting as competitive inhibitors, mutilate the proper binding of substrates through malfunctioning the EPIs movements within cells (Yamaguchi et al., 2015; Sjuts et al., 2016). When EPIs are administered along with the antibiotics, they enhance the retention time of drug through coexistence and maintaining optimum drug concentration in the cell needed for effective treatment of infections due to MDR strains (Coban et al., 2011; Chevalier et al., 2001; Lamut et al., 2019).

4.6.5. Passive Immunization/Therapeutic Antibodies

Various studies around the world have recommended the use of anti-infection immunoglobulins to confront the springing medical problems associated with bacterial infections (Xu et al., 2014). Therapeutic antibodies are in constant use for disease treatment like cancer (Reichert et al., 2005; Trenevska & Banham, 2017). The approach of passive immunization could be traced back to old ages where antisera from inoculated horses were given to the patients for a particular infection (Fernebro, 2011). Yet there is no antibacterial approved therapeutic antibodies (tAbs) except Synagis, which is the only monoclonal antibody marketed for the treatment of respiratory disease (Meulen, 2011).

However, the antibacterial immunoglobulins are categorized into two groups: one group constitutes those candidates which terminate the bacteria through direct binding while the members of another group first disarm the pathogens through removing their virulent fragments and allow the body's immunity to neutralize the invaders (Bebbington & Yarranton, 2008; Fernebro, 2011; Oleksiewicz et al., 2012). Different other tAbs like aurograb, panobacumab and KB001 are in phase II or III clinical trials and showed promising results for certain MDR strains which could provide a natural and economical solution for heath related problems specifically in nosocomial infections (Lu et al., 2011; Amara et al., 2011; Sparrow et al., 2016).

4.6.6. Bacteriophage Therapy

Phage/Bacteriophage is a bacterium specific virus that could parasitize, reproduce and eventually neutralize the respective host (bacteria) through bursting the cell while other phages despite terminating, take control of host through hijacking the genome (O'flaherty et al., 2009). These phages are also known as the natural predators of the various types of bacteria (Chan et al., 2013). In a few US states like Georgia, the population has been exposed to the bacteriophages against several bacterial infections (Kutateladze and Adamia, 2008). Phage therapy was a common practice during the preantibitoic era and abandoned afterward but the emergence of MDR has developed the interest of researchers to resume their use. Two famous phages including the BFC-1 and Biophage-PA have been administered against different infections (Wright et al., 2009; Merabishvili et al., 2009; Capparelli et al., 2010; Xu et al., 2014). Further research is needed to confirm the role and mode of action of phages in combating MDR strains in the future.

4.6.7. Combination Therapy

This novel approach has gained the attention of the world clinician's community for successful management and treatment of MDR Enterobacteriaceae based infections (Trecarichi & Tumbarello, 2017). Now a days, combination therapy has preferred over monotherapy for certain reasons i.e., the synergetic effect of antibiotics, enhanced initial therapeutic activity and possible MDR prevention (Qureshi et al., 2012; Delgado-

Valverde et al., 2013. Different potentially effective combinations are given below:

I. Ceftazidime+Avibactam

It is among the first combinatory drugs proved by the US FDA which contains a member of 3rd generation cephalosporins (Ceftazidime) along with a β-lactamase inhibitor (avibactam) (Wright et al., 2017). Originally, this combination was introduced for the treatment of different infections of urinary tract and pneumonia. Later on, Ceftazidime+avibactamhas been evaluated as a potent remedial agent against Gram-negative bacterial infections (Flamm et al., 2014; Sader et al., 2017; Shirley, 2018). Avibactam protects ceftazidime from hydrolysis by lactamases; ultimately both deliver synergetic effect for MDR strains (Zhanel et al., 2013).

II. Aztreonam+Avibactam

Aztreonam is FDA approved antibiotic since 1980s and has significant antibacterial potential against Enterobactyeriaceae. Moreover, it produces very few allergic reactions as well as does not affect gut flora (Karaiskos et al., 2019). On the other hand, aztreonam alone is readily hydrolyzed by ESBL alone while the combination with avibactam enhances its stability (Yigit et al., 2003). Along with stability, this combination has been proved potent against multiple Gram-negative bacteria like MDR Enterobacteriaceae (Wright et al., 2017).

III. Ceftolozane+Tazobactam

It is a modern cephalosporin (ceftolozane) and penicillin-based β-lactamase inhibitor (tazobactam) combination (Zhanel et al., 2014). Primary studies demonstrate the significant bactericidal *in vitro* activity of this combination against Enterobacteriaceae (Pfaller et al., 2017).

IV. Imipenem+Cilastin+Relebactam

Relebactam is an inhibitor of class A and C beta-lactamases while its combination with imipenem and cilastatin has studied which showed promising antibacterial activity against different strains like *P. aeruginosa* and carbapenemase resistant *E. coli* and other *Enterobacter* species (Lapuebla et al., 2015).

V. Meropenem+Vabrobactam

Vabrobactam is a boronic acid derivative with effective bactericidal potential against class A, C and D beta-lactamases (Bush et al., 2015; Lomovskaya et al., 2017). This drug works by forming a bond between the serine moiety of beta lactamases and its boron atom (Lapuebla et al., 2015). Bactericidal activity against Gram-negative strains like *K. pneumoniae* has been reported when used in combination with meropenem (Livermore et al., 2013; Lomovskaya et al., 2017).

VI. Ceftaroline Fosamil+Avibactam

Alone ceftaroline fosamil, a 5th generation cephalosporin, has wide spectrum bactericidal potential against Gram-positive strains (*Staphylococcus aureus* and *Streptococcus pneumoniae*) and Gram-negative bacteria (non-ESBL producers) (Testa et al., 2015). Avibactam when used in combination with ceftaroline fosamil exhibits significant bactericidal activity for serious infections causing strains including the ESBL and KPC (*K. pneumoniae* carbapenemase) producing like Enterobacteriaceae (Papp-Wallace et al., 2015).

4.6.8. Nanoparticles (A Modern Way to Combat MDR Pathogens)

Nanoparticles (NPs) have been providing a novel and alternate strategy to tackle the infections due to multiple drug-resistant organisms (MDROs) (Singh et al., 2014; Baptista et al., 2018). Harboring the peculiar physical and chemical characters, NPs are expressing the promising therapeutic potentials to overcome health concerns like hurdles in managing the drug resistance (Hemeg et al., 2017; Slavin et al., 2017). In the future, NPs could replace traditional antibiotics due to certain structural and chemical properties like broad surface area which maximizes their contact with pathogens (Lee & Bong-Hyun, 2019). Moreover, nanoparticles also interrupt the normal molecular processes of MDROs as well as outer boundary (cell membrane) disruption (Dakal et al., 2016; Hemeg et al., 2017).

Synergetic bactericidal effect of NPs (Au) with different antibiotics (streptomycin, ampicillin, and kanamycin) against Gram-negative pathogens like MDR Enterobacteriaceae has also been reported (Saha et al., 2007; Gupta et al., 2017). Multiple combinations along with variable proportions of antibiotics and NPs could provide complex bactericidal modes of action; assist to combat the MDR (Gupta et al., 2017). Further deep studies are mandatory to understand the exact mechanisms involve in the combat against the MDR. Although, some published mode of actions is reported as (i) Some NPs mutilate the bacterial cell integrity through membrane lysis (ii) Block the biofilm synthesis (iii) Reactive oxygen species generation (iv) Inhibiting the transcription and translation (Beyth et al., 2015; Lee & Bong-Hyun, 2019). Crisis of the ongoing MDR could be significantly overcome through proper usage of NPs combination therapy (Baptista et al., 2018).

Moreover, combination therapy using different NPs like Ag, Cu, Au, Al, Ni and Zn could also magnify the antimicrobial power of frontline antibiotics including ciprofloxacin, vancomycin, ampicillin and polymyxins against MDR strains like Enterobacteriaceae (Ansari et al., 2014; Hemeg et al., 2017). Last but not the least, options discussed in this chapter to combat MDR infections are reasonable and could be a possible solution to manage the strains through economical and available practices.

References

Abdallah, H. M., Al-Abd, A. M., El-Dine, R. S. & El-Halawany, A. M. (2015). P-glycoprotein inhibitors of natural origin as potential tumor chemo-sensitizers: A review. *Journal of advanced research*, *6*(1), 45-62.

Aeschlimann, J. R., Dresser, L. D., Kaatz, G. W. & Rybak, M. J. (1999). Effects of NorA Inhibitors on *In Vitro* Antibacterial Activities and Postantibiotic Effects of Levofloxacin, Ciprofloxacin, and Norfloxacin in Genetically Related Strains of *Staphylococcus aureus. Antimicrobial agents and chemotherapy*, *43*(2), 335-340.

Aggen, J. B., Armstrong, E. S., Goldblum, A. A., Dozzo, P., Linsell, M. S., Gliedt, M. J. & Lopez, S. (2010). Synthesis and spectrum of the neoglycoside ACHN-490. *Antimicrobial agents and chemotherapy*, *54*(11), 4636-4642.

Aggen, J. B., Armstrong, E. S., Goldblum, A. A., Dozzo, P., Linsell, M. S., Gliedt, M. J. & Lopez, S. (2010). Synthesis and spectrum of the neoglycoside ACHN-490. *Antimicrobial agents and chemotherapy*, *54*(11), 4636-4642.

Aktaş, Z., Kayacan, C. & Oncul, O. (2012).*In vitro* activity of avibactam (NXL104) in combination with β-lactams against Gram-negative bacteria, including OXA-48 β-lactamase-producing *Klebsiella pneumoniae. International journal of antimicrobial agents*, *39*(1), 86-89.

Amara, N., Krom, B. P., Kaufmann, G. F. & Meijler, M. M. (2011). Macromolecular inhibition of quorum sensing: enzymes, antibodies, and beyond. *Chemical Reviews*, *111*(1), 195-208.

Ambler, R. P. (1980).The structure of β-lactamases. *Philosophical Transactions of the Royal Society of London. B, Biological Sciences*, *289*(1036), 321-331.

Ansari, M. A., Khan, H. M., Khan, A. A., Cameotra, S. S., Saquib, Q. & Musarrat, J. (2014). Gum arabic capped-silver nanoparticles inhibit biofilm formation by multidrug resistant strains of *Pseudomonas aeruginosa. Journal of basic microbiology*, *54*(7), 688-699.

Arvizo, R. R., Giri, K., Moyano, D., Miranda, O. R., Madden, B., McCormick, D. J. & Mukherjee, P. (2012). Identifying new therapeutic targets via modulation of protein corona formation by engineered nanoparticles. *PloS one*, *7*(3), e33650.

B Hadley, E. & EW Hancock, R. (2010). Strategies for the discovery and advancement of novel cationic antimicrobial peptides. *Current topics in medicinal chemistry*, *10*(18), 1872-1881.

Baltzer, S. A. & Brown, M. H. (2011). Antimicrobial peptides–promising alternatives to conventional antibiotics. *Journal of molecular microbiology and biotechnology*, *20*(4), 228-235.

Baptista, P. V., McCusker, M. P., Carvalho, A., Ferreira, D. A., Mohan, N. M., Martins, M. & Fernandes, A. R. (2018). Nano-strategies to fight multidrug resistant bacteria—"A Battle of the Titans". *Frontiers in microbiology*, *9*, 1441.

Bartlett, J. G., Gilbert, D. N. & Spellberg, B. (2013). Seven ways to preserve the miracle of antibiotics. *Clinical infectious diseases*, *56*(10), 1445-1450.

Bartlett, J. M. & Siola, P. L. (2014). Implementation and first-year results of an antimicrobial stewardship program at a community hospital. *American journal of health-system pharmacy*, *71*(11), 943-949.

Bassetti, M.& Righi, E. (2014). Eravacycline for the treatment of intra-abdominal infections. *Expert opinion on investigational drugs*, *23*(11), 1575-1584.

Bassetti, M., Righi, E., Fasce, R., Molinari, M. P., Rosso, R., Di Biagio, A. & Viscoli, C. (2007). Efficacy of ertapenem in the treatment of early ventilator-associated pneumonia caused by extended-spectrum β-lactamase-producing organisms in an intensive care unit. *Journal of antimicrobial chemotherapy*, *60*(2), 433-435.

Bbosa, G. S., Mwebaza, N., Odda, J., Kyegombe, D. B. & Ntale, M. (2014). Antibiotics/antibacterial drug use, their marketing and promotion during the post-antibiotic golden age and their role in emergence of bacterial resistance. *Health*, *2014*.

Bebbington, C. & Yarranton, G. (2008). Antibodies for the treatment of bacterial infections: current experience and future prospects. *Current opinion in biotechnology*, *19*(6), 613-619.

Bebrone, C., Lassaux, P., Vercheval, L., Sohier, J. S., Jehaes, A., Sauvage, E. & Galleni, M. (2010). Current challenges in antimicrobial chemotherapy. *Drugs*, *70*(6), 651-679.

Beyth, N., Houri-Haddad, Y., Domb, A., Khan, W. & Hazan, R. (2015). Alternative antimicrobial approach: nano-antimicrobial materials. *Evidence-based complementary and alternative medicine*, *2015*.

Bommarius, B., Jenssen, H., Elliott, M., Kindrachuk, J., Pasupuleti, M., Gieren, H. & Kalman, D. (2010). Cost-effective expression and purification of antimicrobial and host defense peptides in *Escherichia coli*. *Peptides*, *31*(11), 1957-1965.

Bush, K. (2015). A resurgence of β-lactamase inhibitor combinations effective against multidrug-resistant Gram-negative pathogens. *International journal of antimicrobial agents*, *46*(5), 483-493.

Bush, K. (2018). Game changers: new β-lactamase inhibitor combinations targeting antibiotic resistance in Gram-negative bacteria. *ACS infectious diseases*, *4*(2), 84-87.

Bush, K. & Jacoby, G. A. (2010).Updated functional classification of β-lactamases. *Antimicrobial agents and chemotherapy*, *54*(3), 969-976.

Capparelli, R., Nocerino, N., Iannaccone, M., Ercolini, D., Parlato, M., Chiara, M. & Iannelli, D. (2010). Bacteriophage therapy of *Salmonella enterica*: a fresh appraisal of bacteriophage therapy. *The Journal of infectious diseases*, *201*(1), 52-61.

Chan, B. K., Abedon, S. T. & Loc-Carrillo, C. (2013). Phage cocktails and the future of phage therapy. *Future microbiology*, *8*(6), 769-783.

Chevalier, J., Atifi, S., Eyraud, A., Mahamoud, A., Barbe, J. & Pagès, J. M. (2001). New Pyridoquinoline Derivatives as Potential Inhibitors of the Fluoroquinolone Efflux Pump in Resistant *Enterobacter aerogenes* Strains. *Journal of medicinal chemistry*, *44*(23), 4023-4026.

Choi, S. H., Lee, J. E., Park, S. J., Choi, S. H., Lee, S. O., Jeong, J. Y. & Kim, Y. S. (2008). Emergence of antibiotic resistance during therapy for infections caused by Enterobacteriaceae producing AmpC β-lactamase: implications for antibiotic use. *Antimicrobial agents and chemotherapy*, *52*(3), 995-1000.

Coban, A. Y., Guney, A. K., Cayci, Y. T. & Durupinar, B. (2011). Effect of 1-(1-Naphtylmethyl)-piperazine, an efflux pump inhibitor, on antimicrobial drug susceptibilities of clinical *Acinetobacter baumannii* isolates. *Current microbiology*, *62*(2), 508-511.

Collins, V. L., Marchaim, D., Pogue, J. M., Moshos, J., Bheemreddy, S., Sunkara, B. & Blunden, C. (2012). Efficacy of ertapenem for treatment of bloodstream infections caused by extended-spectrum-β-lactamase-producing Enterobacteriaceae. *Antimicrobial agents and chemotherapy*, *56*(4), 2173-2177.

Dakal, T. C., Kumar, A., Majumdar, R. S. & Yadav, V. (2016). Mechanistic basis of antimicrobial actions of silver nanoparticles. *Frontiers in microbiology*, *7*, 1831.

Del Franco, M., Paone, L., Novati, R., Giacomazzi, C. G., Bagattini, M., Galotto, C. & Zarrilli, R. (2015). Molecular epidemiology of carbapenem resistant Enterobacteriaceae in Valle d'Aosta region, Italy, shows the emergence of KPC-2 producing *Klebsiella pneumoniae* clonal complex 101 (ST101 and ST1789). *BMC microbiology*, *15*(1), 260.

Docquier, J. D. & Mangani, S. (2018). An update on β-lactamase inhibitor discovery and development. *Drug Resistance Updates*, *36*, 13-29.

Donowitz, G. R. & Mandell, G. L. (1988). Beta-lactam antibiotics. *New England Journal of Medicine*, *318*(7), 419-426.

Durand-Réville, T. F., Guler, S., Comita-Prevoir, J., Chen, B., Bifulco, N., Huynh, H. & Moussa, S. H. (2017). ETX2514 is a broad-spectrum β-lactamase inhibitor for the treatment of drug-resistant Gram-negative bacteria including *Acinetobacter baumannii*. *Nature microbiology*, *2*(9), 17104.

Ehmann, D. E., Jahić, H., Ross, P. L., Gu, R. F., Hu, J., Kern, G. & Fisher, S. L. (2012). Avibactam is a covalent, reversible, non–β-lactam β-lactamase inhibitor. *Proceedings of the National Academy of Sciences*, *109*(29), 11663-11668.

Falagas, M. E., Karageorgopoulos, D. E. & Nordmann, P. (2011). Therapeutic options for infections with Enterobacteriaceae producing carbapenem-hydrolyzing enzymes. *Future microbiology*, *6*(6), 653-666.

Fernebro, J. (2011). Fighting bacterial infections—future treatment options. *Drug Resistance Updates*, *14*(2), 125-139.

Findlay, B., Zhanel, G. G. & Schweizer, F. (2010). Cationic amphiphiles, a new generation of antimicrobials inspired by the natural antimicrobial peptide scaffold. *Antimicrobial agents and chemotherapy*, *54*(10), 4049-4058.

Fjell, C. D., Hiss, J. A., Hancock, R. E. & Schneider, G. (2012). Designing antimicrobial peptides: form follows function. *Nature reviews Drug discovery*, *11*(1), 37-51.

Flamm, R. K., Stone, G. G., Sader, H. S., Jones, R. N. & Nichols, W. W. (2014). Avibactam reverts the ceftazidime MIC90 of European Gram-negative bacterial clinical isolates to the epidemiological cut-off value. *Journal of Chemotherapy*, *26*(6), 333-338.

Galani, I., Souli, M., Daikos, G. L., Chrysouli, Z., Poulakou, G., Psichogiou, M. & Giamarellou, H. (2012). Activity of plazomicin (ACHN-490) against MDR clinical isolates of *Klebsiella pneumoniae*, *Escherichia coli*, and *Enterobacter* spp. from Athens, Greece. *Journal of Chemotherapy*, *24*(4), 191-194.

Giuliani, A. & Rinaldi, A. C. (2011). Beyond natural antimicrobial peptides: multimeric peptides and other peptidomimetic approaches. *Cellular and Molecular Life Sciences*, *68*(13), 2255-2266.

Giuliani, A., Pirri, G., Bozzi, A., Di Giulio, A., Aschi, M. & Rinaldi, A. C. (2008). Antimicrobial peptides: natural templates for synthetic membrane-active compounds. *Cellular and Molecular Life Sciences*, *65*(16), 2450-2460.

Golkar, Z., Bagasra, O. & Pace, D. G. (2014). Bacteriophage therapy: a potential solution for the antibiotic resistance crisis. *The Journal of Infection in Developing Countries*, *8*(02), 129-136.

Gopalakrishnan, R., Frolov, A. I., Knerr, L., Drury III, W. J. & Valeur, E. (2016). Therapeutic potential of foldamers: from chemical biology tools to drug candidates? *Journal of Medicinal Chemistry*, *59*(21), 9599-9621.

Gould, I. M. &Bal, A. M. (2013). New antibiotic agents in the pipeline and how they can help overcome microbial resistance. *Virulence*, *4*(2), 185-191.

Gross, M. (2013). *Antibiotics in crisis*. Magazine R1063. PMID:24501765. DOI: 10.1016/j.cub.2013.11.057.

Grossman, T. H., Starosta, A. L., Fyfe, C., O'Brien, W., Rothstein, D. M., Mikolajka, A. & Sutcliffe, J. A. (2012). Target-and resistance-based mechanistic studies with TP-434, a novel fluorocycline antibiotic. *Antimicrobial agents and chemotherapy*, *56*(5), 2559-2564.

Gupta, S., Kesarla, R., Chotai, N., Misra, A. & Omri, A. (2017). Systematic approach for the formulation and optimization of solid lipid nanoparticles of efavirenz by high pressure homogenization using design of experiments for brain targeting and enhanced bioavailability. *Biomed research international, 2017.*

Hale, J. D. & Hancock, R. E. (2007). Alternative mechanisms of action of cationic antimicrobial peptides on bacteria. *Expert review of anti-infective therapy*, *5*(6), 951-959.

Hancock, R. E. & Chapple, D. S. (1999). Peptide antibiotics. *Antimicrobial agents and chemotherapy*, *43*(6), 1317-1323.

Harris, P. N. A. & Ferguson, J. K. (2012). Antibiotic therapy for inducible AmpC β-lactamase-producing Gram-negative bacilli: what are the alternatives to carbapenems, quinolones and aminoglycosides? *International journal of antimicrobial agents*, *40*(4), 297-305.

Hemeg, H. A. (2017). Nanomaterials for alternative antibacterial therapy. *International journal of nanomedicine*, *12*, 8211.

Hu, Y., Li, X., Sebti, S. M., Chen, J. &Cai, J. (2011). Design and synthesis of a Apeptides: A new class of peptide mimics. *Bioorganic & medicinal chemistry letters*, *21*(5), 1469-1471.

Ito, A., Nishikawa, T., Matsumoto, S., Yoshizawa, H., Sato, T., Nakamura, R. & Yamano, Y. (2016). Siderophore cephalosporin cefiderocol utilizes ferric iron transporter systems for antibacterial activity against *Pseudomonas aeruginosa. Antimicrobial agents and chemotherapy*, *60*(12), 7396-7401.

Ito, A., Sato, T., Ota, M., Takemura, M., Nishikawa, T., Toba, S. & Nakamura, R. (2018). *In vitro* antibacterial properties of cefiderocol, a novel siderophore cephalosporin, against Gram-negative bacteria. *Antimicrobial agents and chemotherapy*, *62*(1).

Jenssen, H., Hamill, P. & Hancock, R. E. (2006). Peptide antimicrobial agents. *Clinical microbiology reviews*, *19*(3), 491-511.

Jones, R. N., Ferraro, M. J., Reller, L. B., Schreckenberger, P. C., Swenson, J. M. & Sader, H. S. (2007). Multicenter studies of tigecycline disk diffusion susceptibility results for *Acinetobacter* spp. *Journal of clinical microbiology*, *45*(1), 227-230.

Karageorgopoulos, D. E., Wang, R., Yu, X. H. & Falagas, M. E. (2012). Fosfomycin: evaluation of the published evidence on the emergence of antimicrobial resistance in Gram-negative pathogens. *Journal of Antimicrobial Chemotherapy*, *67*(2), 255-268.

Karaiskos, I., Galani, I., Souli, M. & Giamarellou, H. (2019). Novel β-lactam-β-lactamase inhibitor combinations: Expectations for the treatment of carbapenem-resistant Gram-negative pathogens. *Expert Opinion on Drug Metabolism & Toxicology*, *15*(2), 133-149.

Kelly, R. J., Gulley, J. L. & Giaccone, G. (2010). Targeting the immune system in non–small-cell lung cancer: bridging the gap between promising concept and therapeutic reality. *Clinical lung cancer*, *11*(4), 228-237.

Kofteridis, D. P., Alexopoulou, C., Valachis, A., Maraki, S., Dimopoulou, D., Georgopoulos, D. & Samonis, G. (2010). Aerosolized plus intravenous colistin versus intravenous colistin alone for the treatment of ventilator-associated pneumonia: a matched case-control study. *Clinical infectious diseases*, *51*(11), 1238-1244.

Koo, H. B. & Seo, J. (2019). Antimicrobial peptides under clinical investigation. *Peptide Science*, *111*(5), e24122.

Kutateladze, Á. & Adamia, R. (2008). Phage therapy experience at the Eliava Institute. *Médecine et maladies infectieuses*, *38*(8), 426-430.

Lagacé-Wiens, P. R. S., Tailor, F., Simner, P., DeCorby, M., Karlowsky, J. A., Walkty, A. & Zhanel, G. G. (2011). Activity of NXL104 in combination with β-lactams against genetically characterized Escherichia coli and *Klebsiella pneumoniae* isolates producing class A extended-spectrum β-lactamases and class C β-lactamases. *Antimicrobial agents and chemotherapy*, *55*(5), 2434-2437.

Lahiri, S. D., Bradford, P. A., Nichols, W. W. & Alm, R. A. (2016). Structural and sequence analysis of class A β-lactamases with respect to avibactam inhibition: impact of Ω-loop variations. *Journal of Antimicrobial Chemotherapy*, *71*(10), 2848-2855.

Lamut, A., Peterlin Mašič, L., Kikelj, D. & Tomašič, T. (2019). Efflux pump inhibitors of clinically relevant multidrug resistant bacteria. *Medicinal Research Reviews*, *39*(6), 2460-2504.

Landman, D., Georgescu, C., Martin, D. A. & Quale, J. (2008). Polymyxins revisited. *Clinical microbiology reviews*, *21*(3), 449-465.

Lapuebla, A., Abdallah, M., Olafisoye, O., Cortes, C., Urban, C., Landman, D. & Quale, J. (2015). Activity of imipenem with relebactam against Gram-negative pathogens from New York City. *Antimicrobial agents and chemotherapy*, *59*(8), 5029-5031.

Lee, S. H. & Jun, B. H. (2019). Silver nanoparticles: synthesis and application for nanomedicine. *International journal of molecular sciences*, *20*(4), 865.

Levy, S. B. & Marshall, B. (2004). Antibacterial resistance worldwide: causes, challenges and responses. *Nature medicine*, *10*(12), S122-S129.

Lin, J., Nishino, K., Roberts, M. C., Tolmasky, M., Aminov, R. I. & Zhang, L. (2015). Mechanisms of antibiotic resistance. *Frontiers in microbiology*, *6*, 34.

Lipsky, B. A., Holroyd, K. J. & Zasloff, M. (2008). Topical versus systemic antimicrobial therapy for treating mildly infected diabetic foot ulcers: a randomized, controlled, double-blinded, multicenter trial of pexiganan cream. *Clinical infectious diseases*, *47*(12), 1537-1545.

Liu, C. Z., Koppireddi, S., Wang, H., Zhang, D. W. & Li, Z. T. (2019). Halogen Bonding Directed Supramolecular Quadruple and Double Helices from Hydrogen-Bonded Arylamide Foldamers. *Angewandte Chemie*, *131*(1), 232-236.

Livermore, D. M., Mushtaq, S., Barker, K., Hope, R., Warner, M. & Woodford, N. (2012). Characterization of β-lactamase and porin mutants of Enterobacteriaceae selected with ceftaroline+ avibactam (NXL104). *Journal of antimicrobial chemotherapy*, *67*(6), 1354-1358.

Livermore, D. M., Mushtaq, S., Warner, M. & Woodford, N. (2016). *In vitro* activity of eravacycline against carbapenem-resistant Enterobacteriaceae and *Acinetobacter baumannii*. *Antimicrobial agents and chemotherapy*, *60*(6), 3840-3844.

Livermore, D. M., Mushtaq, S., Warner, M., Zhang, J. C., Maharjan, S., Doumith, M. & Woodford, N. (2011). Activity of aminoglycosides, including ACHN-490, against carbapenem-resistant Enterobacteriaceae isolates. *Journal of Antimicrobial Chemotherapy*, *66*(1), 48-53.

Lomovskaya, O., Sun, D., Rubio-Aparicio, D., Nelson, K., Tsivkovski, R., Griffith, D. C. & Dudley, M. N. (2017). Vaborbactam: spectrum of beta-lactamase inhibition and impact of resistance mechanisms on activity in Enterobacteriaceae. *Antimicrobial agents and chemotherapy*, *61*(11).

López-Cerero, L., Picón, E., Morillo, C., Hernández, J. R., Docobo, F., Pachón, J. & Pascual, A. (2010). Comparative assessment of inoculum effects on the antimicrobial activity of amoxycillin-clavulanate and piperacillin-tazobactam with extended-spectrum β-lactamase-producing and extended-spectrum β-lactamase-non-producing Escherichia coli isolates. *Clinical microbiology and infection*, *16*(2), 132-136.

Louie, A., Castanheira, M., Liu, W., Grasso, C., Jones, R. N., Williams, G. & Kulawy, R. (2012). Pharmacodynamics of β-lactamase inhibition by NXL104 in combination with ceftaroline: examining organisms with multiple types of β-lactamases. *Antimicrobial agents and chemotherapy*, *56*(1), 258-270.

Lu, Q., Rouby, J. J., Laterre, P. F., Eggimann, P., Dugard, A., Giamarellos-Bourboulis, E. J. & Georgescu-Kyburz, V. (2011). Pharmacokinetics and safety of panobacumab: specific adjunctive immunotherapy in critical patients with nosocomial *Pseudomonas aeruginosa* O11 pneumonia. *Journal of antimicrobial chemotherapy*, *66*(5), 1110-1116.

Lushniak, B. D. (2014). Antibiotic resistance: a public health crisis. *Public Health Reports*, *129*(4), 314-316.

Luyt, C. E., Bréchot, N., Trouillet, J. L. & Chastre, J. (2014). Antibiotic stewardship in the intensive care unit. *Critical care*, *18*(5), 480.

Marc, C., Vrignaud, B., Levieux, K., Robine, A., Guen, C. G. L. & Launay, E. (2016). Inappropriate prescription of antibiotics in pediatric practice: analysis of the prescriptions in primary care. *Journal of Child Health Care*, *20*(4), 530-536.

Melo, M. N., Dugourd, D. & Castanho, M. A. (2006). Omiganan pentahydrochloride in the front line of clinical applications of antimicrobial peptides. *Recent patents on anti-infective drug discovery*, *1*(2), 201-207.

Mensa, B., Kim, Y. H., Choi, S., Scott, R., Caputo, G. A. & DeGrado, W. F. (2011). Antibacterial mechanism of action of arylamide foldamers. *Antimicrobial agents and chemotherapy*, *55*(11), 5043-5053.

Merabishvili, M., Pirnay, J. P., Verbeken, G., Chanishvili, N., Tediashvili, M., Lashkhi, N. & Lavigne, R. (2009). Quality-controlled small-scale production of a well-defined bacteriophage cocktail for use in human clinical trials. *PloS one*, *4*(3), e4944.

Michael, C. A., Dominey-Howes, D. & Labbate, M. (2014). The antimicrobial resistance crisis: causes, consequences, and management. *Frontiers in public health*, *2*, 145.

Michalopoulos, A. S., Livaditis, I. G. & Gougoutas, V. (2011). The revival of fosfomycin. *International journal of infectious diseases*, *15*(11), e732-e739.

Michalopoulos, A., Virtzili, S., Rafailidis, P., Chalevelakis, G., Damala, M. & Falagas, M. E. (2010). Intravenous fosfomycin for the treatment of nosocomial infections caused by carbapenem-resistant *Klebsiella pneumoniae* in critically ill patients: a prospective evaluation. *Clinical microbiology and infection*, *16*(2), 184-186.

Milatovic, D., Schmitz, F. J., Verhoef, J. & Fluit, A. C. (2003). Activities of the glycylcycline tigecycline (GAR-936) against 1,924 recent European clinical bacterial isolates. *Antimicrobial agents and chemotherapy*, *47*(1), 400-404.

Molchanova, N., Hansen, P. R. & Franzyk, H. (2017). Advances in development of antimicrobial peptidomimetics as potential drugs. *Molecules*, *22*(9), 1430.

Navarro, F., Miro, E. & Mirelis, B. (2010). Interpretive reading of enterobacteria antibiograms. *Enfermedades infecciosas y microbiologia clinica*, *28*(9), 638-645.

Nicolau, D. P., Carmeli, Y., Crank, C. W., Goff, D. A., Graber, C. J., Lima, A. L. L. & Goldstein, E. J. (2012). Carbapenem stewardship: does ertapenem affect Pseudomonas susceptibility to other carbapenems? A review of the evidence. *International journal of antimicrobial agents*, *39*(1), 11-15.

Nikaido, H. (2011). Structure and mechanism of RND-type multidrug efflux pumps. *Advances in enzymology and related areas of molecular biology*, *77*, 1.

Noskin, G. A. (2005). Tigecycline: a new glycylcycline for treatment of serious infections. *Clinical infectious diseases*, *41*(Supplement_5), S303-S314.

O'Flaherty, S., Ross, R. P. & Coffey, A. (2009). Bacteriophage and their lysins for elimination of infectious bacteria. *FEMS microbiology reviews*, *33*(4), 801-819.

Oleksiewicz, M. B., Nagy, G. & Nagy, E. (2012). Anti-bacterial monoclonal antibodies: back to the future? *Archives of biochemistry and biophysics*, *526*(2), 124-131.

Padhee, S., Hu, Y., Niu, Y., Bai, G., Wu, H., Costanza, F. & Cai, J. (2011). Non-hemolytic α-AApeptides as antimicrobial peptidomimetics. *Chemical communications*, *47*(34), 9729-9731.

Papp-Wallace, K. M., Bajaksouzian, S., Abdelhamed, A. M., Foster, A. N., Winkler, M. L., Gatta, J. A. & Jacobs, M. R. (2015). Activities of ceftazidime, ceftaroline, and aztreonam alone and combined with avibactam against isogenic *Escherichia coli* strains expressing selected single β-lactamases. *Diagnostic microbiology and infectious disease*, *82*(1), 65-69.

Papp-Wallace, K. M., Endimiani, A., Taracila, M. A. & Bonomo, R. A. (2011). Carbapenems: past, present, and future. *Antimicrobial agents and chemotherapy*, *55*(11), 4943-4960.

Paterson, D. L. (2006). Resistance in gram-negative bacteria: Enterobacteriaceae. *American journal of infection control*, *34*(5), S20-S28.

Peters, A. D., Borsley, S., della Sala, F., Cairns-Gibson, D. F., Leonidou, M., Clayden, J. & Cockroft, S. L. (2020). Switchable foldamer ion channels with antibacterial activity. *Chemical Science*.

Pfaller, M. A., Bassetti, M., Duncan, L. R. & Castanheira, M. (2017). Ceftolozane/tazobactam activity against drug-resistant Enterobacteriaceae and Pseudomonas aeruginosa causing urinary tract and intraabdominal infections in Europe: report from an antimicrobial surveillance programme (2012–15). *Journal of Antimicrobial Chemotherapy*, *72*(5), 1386-1395.

Piddock, L. J. (2012). The crisis of no new antibiotics—what is the way forward? *The Lancet infectious diseases*, *12*(3), 249-253.

Potter, R. F., D'Souza, A. W. & Dantas, G. (2016). The rapid spread of carbapenem-resistant Enterobacteriaceae. *Drug Resistance Updates*, *29*,30-46.

Powers, J. P. S. & Hancock, R. E. (2003). The relationship between peptide structure and antibacterial activity. *Peptides*, *24*(11), 1681-1691.

Queenan, A. M., Shang, W., Kania, M., Page, M. G. & Bush, K. (2007). Interactions of ceftobiprole with β-lactamases from molecular classes A to D. *Antimicrobial agents and chemotherapy*, *51*(9), 3089-3095.

Qureshi, Z. A., Paterson, D. L., Potoski, B. A., Kilayko, M. C., Sandovsky, G., Sordillo, E. & Doi, Y. (2012). Treatment outcome of bacteremia due to KPC-producing *Klebsiella pneumoniae*: superiority of combination antimicrobial regimens. *Antimicrobial agents and chemotherapy*, *56*(4), 2108-2113.

Read, A. F. & Woods, R. J. (2014). Antibiotic resistance management. *Evolution, medicine, and public health*, *2014*(1), 147.

Reichert, J. M., Rosensweig, C. J., Faden, L. B. & Dewitz, M. C. (2005). Monoclonal antibody successes in the clinic. *Nature biotechnology*, *23*(9), 1073-1078.

Rossolini, G. M., Arena, F., Pecile, P. & Pollini, S. (2014). Update on the antibiotic resistance crisis. *Current opinion in pharmacology*, *18*, 56-60.

Roux, K. J., Kim, D. I., Raida, M. & Burke, B. (2012). A promiscuous biotin ligase fusion protein identifies proximal and interacting proteins in mammalian cells. *The Journal of cell biology*, *196*(6), 801-810.

Rubinchik, E., Dugourd, D., Algara, T., Pasetka, C. & Friedland, H. D. (2009). Antimicrobial and antifungal activities of a novel cationic antimicrobial peptide, omiganan, in experimental skin colonisation models. *International journal of antimicrobial agents*, *34*(5), 457-461.

Ruzin, A., Petersen, P. J. & Jones, C. H. (2010). Resistance development profiling of piperacillin in combination with the novel β-lactamase inhibitor BLI-489. *Journal of antimicrobial chemotherapy*, *65*(2), 252-257.

Sader, H. S., Rhomberg, P. R., Flamm, R. K., Jones, R. N. & Castanheira, M. (2017). WCK 5222 (cefepime/zidebactam) antimicrobial activity tested against Gram-negative organisms producing clinically relevant β-lactamases. *Journal of Antimicrobial Chemotherapy*, *72*(6), 1696-1703.

Saha, B., Bhattacharya, J., Mukherjee, A., Ghosh, A., Santra, C., Dasgupta, A. K. & Karmakar, P. (2007). *In vitro* structural and functional evaluation of gold nanoparticles conjugated antibiotics. *Nanoscale Research Letters*, *2*(12), 614.

Saha, M., Sarkar, S., Sarkar, B., Sharma, B. K., Bhattacharjee, S. & Tribedi, P. (2016). Microbial siderophores and their potential applications: a review. *Environmental Science and Pollution Research*, *23*(5), 3984-3999.

Santajit, S. & Indrawattana, N. (2016). Mechanisms of antimicrobial resistance in ESKAPE pathogens. *BioMed research international*, 2016.

Sanyasi, S., Majhi, R. K., Kumar, S., Mishra, M., Ghosh, A., Suar, M., ...& Goswami, L. (2016). Polysaccharide-capped silver Nanoparticles inhibit biofilm formation and eliminate multi-drug-resistant bacteria by disrupting bacterial cytoskeleton with reduced cytotoxicity towards mammalian cells. *Scientific reports*, *6*(1), 1-16.

Scorciapino, M. A., Pirri, G., Vargiu, A. V., Ruggerone, P., Giuliani, A., Casu, M. & Rinaldi, A. C. (2012). A novel dendrimeric peptide with antimicrobial properties: structure-function analysis of SB056. *Biophysical journal*, *102*(5), 1039-1048.

Shirley, M. (2018). Ceftazidime-avibactam: a review in the treatment of serious gram-negative bacterial infections. *Drugs*, *78*(6), 675-692.

Singh, R., Smitha, M. S. & Singh, S. P. (2014). The role of nanotechnology in combating multi-drug resistant bacteria. *Journal of nanoscience and nanotechnology*, *14*(7), 4745-4756.

Sjuts, H., Vargiu, A. V., Kwasny, S. M., Nguyen, S. T., Kim, H. S., Ding, X. & Pos, K. M. (2016). Molecular basis for inhibition of AcrB multidrug efflux pump by novel and powerful pyranopyridine derivatives. *Proceedings of the National Academy of Sciences*, *113*(13), 3509-3514.

Skurnik, D., Kropec, A., Roux, D., Theilacker, C., Huebner, J. & Pier, G. B. (2012). Natural antibodies in normal human serum inhibit *Staphylococcus aureus* capsular polysaccharide vaccine efficacy. *Clinical infectious diseases*, *55*(9), 1188-1197.

Slavin, Y. N., Asnis, J., Häfeli, U. O. & Bach, H. (2017). Metal nanoparticles: understanding the mechanisms behind antibacterial activity. *Journal of nanobiotechnology*, *15*(1), 1-20.

Sparrow, E., Friede, M., Sheikh, M., Torvaldsen, S. & Newall, A. T. (2016). Passive immunization for influenza through antibody therapies, a review of the pipeline, challenges and potential applications. *Vaccine*, *34*(45), 5442-5448.

Spellberg, B. & Gilbert, D. N. (2014). The future of antibiotics and resistance: a tribute to a career of leadership by John Bartlett. *Clinical infectious diseases*, *59*(suppl_2), S71-S75.

Sutcliffe, J. A., O'Brien, W., Fyfe, C. & Grossman, T. H. (2013). Antibacterial activity of eravacycline (TP-434), a novel fluorocycline, against hospital and community pathogens. *Antimicrobial agents and chemotherapy*, *57*(11), 5548-5558.

Tacconelli, E., Carrara, E., Savoldi, A., Harbarth, S., Mendelson, M., Monnet, D. L. & Ouellette, M. (2018). Discovery, research, and development of new antibiotics: the WHO priority list of antibiotic-resistant bacteria and tuberculosis. *The Lancet Infectious Diseases*, *18*(3), 318-327.

ter Meulen, J. (2011). Monoclonal antibodies in infectious diseases: clinical pipeline in 2011. *Infectious Disease Clinics*, *25*(4), 789-802.

Testa, R., Cantón, R., Giani, T., Morosini, M. I., Nichols, W. W., Seifert, H. & Nordmann, P. (2015). *In vitro* activity of ceftazidime, ceftaroline and aztreonam alone and in combination with avibactam against European Gram-negative and Gram-positive clinical isolates. *International journal of antimicrobial agents*, *45*(6), 641-646.

Tew, G. N., Scott, R. W., Klein, M. L. & DeGrado, W. F. (2010). De novo design of antimicrobial polymers, foldamers, and small molecules: from discovery to practical applications. *Accounts of chemical research*, *43*(1), 30-39.

Thaker, H. D., Sgolastra, F., Clements, D., Scott, R. W. & Tew, G. N. (2011). Synthetic mimics of antimicrobial peptides from triaryl scaffolds. *Journal of medicinal chemistry*, *54*(7), 2241-2254.

Trecarichi, E. M. & Tumbarello, M. (2017). Therapeutic options for carbapenem-resistant Enterobacteriaceae infections. *Virulence*, *8*(4), 470-484.

Trenevska, I., Li, D. & Banham, A. H. (2017). Therapeutic antibodies against intracellular tumor antigens. *Frontiers in immunology*, *8*, 1001.

Vardakas, K. Z., Tansarli, G. S., Rafailidis, P. I. & Falagas, M. E. (2012). Carbapenems versus alternative antibiotics for the treatment of bacteraemia due to Enterobacteriaceae producing extended-spectrum β-lactamases: a systematic review and meta-analysis. *Journal of antimicrobial chemotherapy*, *67*(12), 2793-2803.

Velkov, T., Thompson, P. E., Nation, R. L. & Li, J. (2010). Structure–activity relationships of polymyxin antibiotics. *Journal of medicinal chemistry*, *53*(5), 1898-1916.

Venkatesh, M. P. & Rong, L. (2008). Human recombinant lactoferrin acts synergistically with antimicrobials commonly used in neonatal practice against coagulase-negative *Staphylococci* and *Candida albicans* causing neonatal sepsis. *Journal of medical microbiology*, *57*(9), 1113-1121.

Ventola, C. L. (2015). The antibiotic resistance crisis: part 1: causes and threats. *Pharmacy and therapeutics*, *40*(4), 277.

Venugopal, A. A. & Johnson, S. (2012). Fidaxomicin: a novel macrocyclic antibiotic approved for treatment of *Clostridium difficile* infection. *Clinical infectious diseases*, *54*(4), 568-574.

Viswanathan, R., Singh, A. K., Basu, S., Chatterjee, S., Roy, S. & Isaacs, D. (2014). Multi-drug-resistant, non-fermenting, gram-negative bacilli in neonatal sepsis in Kolkata, India: a 4-year study. *Paediatrics and international child health*, *34*(1), 56-59.

Wittebole, X., De Roock, S. & Opal, S. M. (2014). A historical overview of bacteriophage therapy as an alternative to antibiotics for the treatment of bacterial pathogens. *Virulence*, *5*(1), 226-235.

Wright, A., Hawkins, C. H., Änggård, E. E. & Harper, D. R. (2009). A controlled clinical trial of a therapeutic bacteriophage preparation in chronic otitis due to antibiotic-resistant *Pseudomonas aeruginosa*; a preliminary report of efficacy. *Clinical otolaryngology*, *34*(4), 349-357.

Wright, H., Bonomo, R. A. & Paterson, D. L. (2017). New agents for the treatment of infections with Gram-negative bacteria: restoring the miracle or false dawn? *Clinical Microbiology and Infection*, *23*(10), 704-712.

Xiao, X. Y., Hunt, D. K., Zhou, J., Clark, R. B., Dunwoody, N., Fyfe, C. & Sun, C. (2012). Fluorocyclines. 1. 7-fluoro-9-pyrrolidinoacetamido-6-demethyl-6-deoxytetracycline: a potent, broad spectrum antibacterial agent. *Journal of medicinal chemistry*, *55*(2), 597-605.

Xu, Z. Q., Flavin, M. T. & Flavin, J. (2014). Combating multidrug-resistant Gram-negative bacterial infections. *Expert opinion on investigational drugs*, *23*(2), 163-182.

Xu, Z. Q., Flavin, M. T. & Flavin, J. (2014). Combating multidrug-resistant Gram-negative bacterial infections. *Expert opinion on investigational drugs*, *23*(2), 163-182.

Y Mahmood, H., Jamshidi, S., Mark Sutton, J. & M Rahman, K. (2016). Current advances in developing inhibitors of bacterial multidrug efflux pumps. *Current Medicinal Chemistry*, *23*(10), 1062-1081.

Yamaguchi, A., Nakashima, R. & Sakurai, K. (2015). Structural basis of RND-type multidrug exporters. *Frontiers in microbiology*, *6*, 327.

Yashima, E., Ousaka, N., Taura, D., Shimomura, K., Ikai, T. & Maeda, K. (2016). Supramolecular helical systems: helical assemblies of small molecules, foldamers, and polymers with chiral amplification and their functions. *Chemical Reviews*, *116*(22), 13752-13990.

Yigit, H., Queenan, A. M., Rasheed, J. K., Biddle, J. W., Domenech-Sanchez, A., Alberti, S. & Tenover, F. C. (2003). Carbapenem-resistant strain of *Klebsiella oxytoca* harboring carbapenem-hydrolyzingβ-lactamase KPC-2. *Antimicrobial agents and chemotherapy*, *47*(12), 3881-3889.

Yin, Y., Song, T., Liao, B., Luo, Q. & Zhou, Z. (2012). Antibiotic prophylaxis in patients undergoing open mesh repair of inguinal hernia: a meta-analysis. *The American Surgeon*, *78*(3), 359-365.

Zhanel, G. G., Cheung, D., Adam, H., Zelenitsky, S., Golden, A., Schweizer, F. & Hoban, D. J. (2016). Review of eravacycline, a novel fluorocycline antibacterial agent. *Drugs*, *76*(5), 567-588.

Zhanel, G. G., Lawson, C. D., Adam, H., Schweizer, F., Zelenitsky, S., Lagacé-Wiens, P. R. & Lynch, J. P. (2013). *Ceftazidime-avibactam*: a novel cephalosporin/β-lactamase inhibitor combination. *Drugs*, *73*(2), 159-177.

Zhanel, G. G., Lawson, C. D., Zelenitsky, S., Findlay, B., Schweizer, F., Adam, H. & Lynch, J. P. (2012). Comparison of the next-generation aminoglycoside plazomicin to gentamicin, tobramycin and amikacin. *Expert review of anti-infective therapy*,*10*(4), 459-473.

Zhanel, G. G., Sniezek, G., Schweizer, F., Zelenitsky, S., Lagace-Wiens, P. R., Rubinstein, E. & Karlowsky, J. A. (2009). Ceftaroline. *Drugs*, *69*(7), 809-831.

In: Recent Developments in *Enterobacter* ... ISBN: 978-1-53618-615-4
Editor: Grégoire Roux

Chapter 4

MULTI-DRUG RESISTANT *ENTEROBACTER CLOACAE* COMPLEX: A GLOBAL DIVERSIFYING THREAT

***Maira Ashraf*[1]*, Muhammad Asad*[2] *and Safdar Ali Mirza*[1,*]**
[1]Department of Botany, GCU, Lahore, Punjab, Pakistan
[2]Central Park Medical College, Lahore, Punjab, Pakistan

ABSTRACT

The *Enterobacter cloacae* Complex (ECC) as a human pathogen has wide geological spread causing infections of urinary tract, respiratory tract, intra-abdominal and bacteremia. Numerous lineages show multiple resistance traits like Ampicillin C (AmpC) and clonal lineages of ECC with high epidemic potential can be linked with antibiotic resistance evolution. Some examples of antibiotic resistant traits in ECC are Extended-spectrum β-lactamases (ESBLs), Metallo-β-lactamases (MBLs), Cephalosporinases and Carbapenemases. In hospital environment, surfaces transmit several key healthcare-associated

* Corresponding Author's Email: safdaralimirza@gcu.edu.pk.

pathogens including ECC. Other factors causing multi-drug resistance include under or overuse of antibiotics, prolonged use of antibiotics, poor infection control, poor hygiene and sanitation. Multi-drug resistance of ECC has become a major challenge for physicians as it complicates treatment of infections; many people appear non-compliant with full regime of treatment due to development of drug resistance in bacteria. Molecular studies reveal that drug resistance has intrinsic (chromosomal) and extrinsic (plasmid) mutation mechanisms. Mutated resistance carrying bacterial strains survive and replicate evolving into the dominant types. Resistance against broad spectrum cephalosporins and aztreonam is apparently caused by hyper production of constitutive AmpC due to multiple regulatory mutations. Multiple resistant genes for production of ESBLs and MBLs and quinolone resistance-determining region (QRDR) gene mutations have been identified to cause resistance against fluoroquinolones. There is a major threat that multi-drug resistance in ECC can lead the world to a post-antibiotic era where infection or injury can result into non-treatable or fatal complications and it can have a drastic impact on the global economy through increase in disease burden as well.

Keywords: *Enterobacter cloacae* complex, multiple resistance traits, antibiotic resistance, quinolone resistance-determining regions

1. Introduction

Enterobacter cloacae complex belonging to family Enterobacteriaceae includes gram-negative, saprophytic and facultative bacteria mostly found in soil and sewage and part of commensal enteric flora of human gastrointestinal tract. *Enterobacter cloacae* complex has wide geological spread and emerged as nosocomial pathogen from intensive care centers producing infection in debilitated patients if they are kept in hospitals for long time and upon prolonged antibiotics treatment thus complicating its therapy (Izdebski et al., 2015). Among twenty-two species of genus *Enterobacter*, ECC includes seven species viz *Enterobacter cloacae*, *Enterobacter asburiae*, *Enterobacter hormaechei*, *Enterobacter kobei*, *Enterobacter ludwigii,* *Enterobacter mori* and *Enterobacter nimipressuralis*, because these have 61%-67% of DNA relatedness with *E.*

cloacae. Enterobacter cloacae and *E. hormaechei* are mostly found causing infection in human, *Enterobacter cloacae* being the major risk globally accounting for up to 10% of postsurgical peritonitis cases, 5% of nosocomial pneumonias, 5% of hospital-acquired sepsis and 4% of nosocomial urinary tract infections. These ECC strains are well-adapted to the hospital environment and can acquire the virulence in form of resistance genes through genetic mobile elements thus possessing multi-drug resistance traits (Mezzatesta et al., 2012). Harbarth et al. (1999) reported the outbreak of *Enterobacter cloacae* in overcrowded and poor hygienic areas. Plasmid-mediated horizontal gene transfer by the movement of resistance genes among different bacterial species results in development of multi-drug resistance potential in bacterial populations. Ren et al. (2010) and Humann et al. (2011) working on genome sequence of *Enterobacter cloacae* and Liu et al. (2016) on *E. asburiae* described that many drug resistance genes are present in these organisms. Due to development of multi-drug resistance in ECC most of the treatable bacteria are now untreatable or need last line of antibiotics (Al-Tawfifiq et al., 2017). Multi-drug resistant clones containing plasmid harboring genes resistant to antibacterial agents are spreading worldwide rapidly and multi-drug resistance is the emerging challenge for modern medicine. Pandemic clones of ECC are the major risk for global public health. To prevent the spread of multi-drug resistant clones of ECC, their surveillance is needed at hospital as well as at community level (Brkic, et al., 2019).

2. Clinical Importance of ECC

Enterobacter cloacae complex is among the most resistant pathogens responsible for nosocomial infections including urinary tract infections, respiratory tract infections, bacteremia, endocarditis, septic arthritis, osteomyelitis, ophthalmic infections, skin and soft tissue infections and intra-abdominal infections in humans especially in elderly and vulnerable sick ICU patients (Kim et al., 2012). *Enterobacter cloacae* complex is a broad spectrum β-lactamase and carbapenemase-producing pathogen causing opportunistic infections in immunocompromised persons causing morbidity and mortality in the world (Liu et al., 2004). *Enterobacter cloacae* is commonly contracted through skin and gastrointestinal tract and found nosocomial pathogen in neonatal units. It also leads to biofilm-related infections like biliary tract infections and catheter-associated urinary tract infections (Fernandez-Baca et al., 2001). *Enterobacter asburiae* is mostly found to be associated with gastrointestinal infections (Brenner et al., 1986), *E. hormaechei* with bloodstream infections (O'Hara et al., 1989), *E. kobei* with nosocomial urosepsis (Hoffmann et al., 2005a), *E. nimipressuralis* with pseudobacteremia (Kim et al., 2010) and *E. ludwigii* with urinary tract infections (Hoffmann et al., 2005b). The clinical importance of ECC marks it dangerous microorganism for modern medicine due to its ability to acquire and upregulate resistance traits.

3. Development of Drug Resistance in ECC and Its Spread

Major factor in emergence of multi-drug resistance in ECC is the misuse and overuse of antibiotics (Jean and Hsueh, 2011). Many patients use antibiotics as self-medication often accompanied by overuse, such uncontrolled use of broad-spectrum antibiotics results in emergence of resistance especially in countries with poor health-care system (Van-Boeckel et al., 2014). Another factor causing drug resistance in ECC may be underuse of medicine thus making the bacteria resistant, these bacteria can spread in other humans (WHO, 2018). According to Brkic et al. (2019) overuse or misuse of carbapenem for treating urinary tract infections in primary healthcare centers in Serbia and multi-drug resistant phenotypes associated with production of bla_{NDM} may be the reasons for development of carbapenem resistance in ECC

Consumption of inadequately cooked or contaminated food with antibiotic resistant ECC, supports drug resistance. Multi-drug resistant bacteria are already present in environment including food items, it spreads the multi-drug resistant ECC in the food chain to humans (Esteban-Cuesta et al., 2019). Seafood also contributes antibiotic resistant genes in the local non-resistant ECC strains. Some ECC strains acquire resistance against aztreonam, carbapenems and penicillins from bacterial strains present in seafood (Brouwer et al., 2019). Some carbapenemase-producing genes were found in both non-pathogenic and pathogenic aquatic bacterial species isolated from seafood. The plasmids and chromosomal regions carrying antibiotic resistant integrated genes cause spread in pathogenic and commensal local bacterial strains of ECC.

The veins of nasal mucosa are more likely to be infected by bacteria as the extensive use of nasal catheter can destroy the normal barrier leading to bacterial translocation. Carbapenem-resistant ECC can colonize on catheter and enter the body in blood circulation. The patients with venous catheterization possess poor immunity and normally are in serious conditions making them more likely to be infected by drug resistant ECC (Soi et al., 2016).

Animals like urban brown rats (*Rattus norvegicus*) in Vienna, Austria have been found to play a role in distribution of antimicrobial resistant genes such as multi-drug resistance in Enterobacteriaceae, especially ST_{114} clones of ECC due to the widespread existence of the rats in the city. Two cefotaxime-resistant members of ECC were isolated from intestines of rats, these isolates showed resistance against ampicillin, β-lactams, cefazoline, tetracycline, fluoroquinolones, carbapenems but were susceptible to amikacin (Desvars-Larrive et al., 2019).

4. Mechanisms of Resistance

To resist the activity of antibiotics, ECC possesses or acquires a variety of different mechanisms. There are many physical and chemical modifications that are adapted by ECC for development of antibiotic resistance. Physical modifications include development of membrane transporters like efflux pumps or making the membrane impermeable for antibiotics. *Enterobacter cloacae* Complex, being gram-negative bacteria, possess outer membrane and antibiotics have to cross the outer membrane to interact with the target in metabolism thus mutations leading to alterations or loss of outer-membrane channels known as porins results in development of drug resistance (Fernandez and Hancock, 2012). Physical modifications are basically due to genetic modification or alteration in certain genes. In a research Majewski et al. (2016) concluded that moderate to high level carbapenem resistance was due to decreased porin permeability in many isolates of ECC and Wozniak et al. (2012) described that some isolates showed intermediate or high level of resistance against carbapenems without production of carbapenemase enzymes. The resistant mechanism used by these isolates was alteration of some porins coupled with overexpression of *ampC* genes. Huang et al. (2019) found the mechanism for colistin heteroresistance very complicated involving multiple genes. For the first time, it was found that inner membrane protein-encoding genes i.e., *phoP*, *dedA(Ecl)*, *tolC* and AcrAB-TolC efflux pump causes colistin heteroresistance in ECC.

Chemical modifications include the production of certain chemicals i.e., enzymes that can easily hydrolyze the antibiotic used. Chemical modifications are caused by various regulator genes, genetic alterations and protein expression by production of different enzymes. For instance, production of β-lactamases is responsible for resistance against β-lactam (Davin-Regli and Pages, 2015), AmpC β-lactamase-type cephalosporinase produced resistance against first generation cephalosporins (Carter et al., 2017), resistance against carbapenem is induced by production of carbapenemases (Peirano et al., 2018). Regulatory pathways followed by ECC for AmpC-mediated β-lactam resistance is reported by Guerin et al. (2015). Genetic modification can be both chromosomal or plasmid-mediated. Repression of promoter region of chromosomal-mediated *ampC* gene make them resistant to ampicillin, cefuroxime and cefazolin by low level production of AmpC enzymes. Resistance against broad-spectrum cephalosporins including ceftazidime, ceftriaxone and cefotaxime can be induced by derepressed mutations causing high level production of AmpC enzymes. Overproduction of AmpC β-lactamases can be induced by derepression of chromosomal genes or by acquisition of transfer-derepressed plasmid (Jacoby, 2009). Co-existence of ESBL genes with *bla*$_{DNM-1}$, *bla*$_{IMP-26}$ and *bla*$_{IMP-8}$ genes confered carbapenem resistance in some ECC strains in China (Huang et al., 2012). Development of antibiotic resistance, regulation of drug resistance genes and its clinical importance need to be focused as innovative research.

5. Antibiotic Resistance of ECC

Many isolates of ECC are resistant to ampicillin, amoxicillin, flouroquinolones, tetracyclines, trimethoprim/ sulfamethoxazole, aminoglycosides, chloramphenicol, first generation cephalosporins, amoxicillin-clavulanate, piperacillin-tazobactam, cefoxitin and carbapenems owing to showing various resistant traits like AmpC, β-lactamases, carbapenemases, cephalosporinases etc. while other are susceptible to some of these drugs. *Enterobacter hormaechei* is susceptible to fosfomycin but *E. cloacae* and *E. asburiae* are both resistant as well as sensitive to it. *Enterobacter asburiae* is naturally resistant to cephalosporins but susceptible to carbapenems, some strains of *E. asburiae* are also resistant to cefalotin, tetracycline and ampicillin (Brenner et al., 1986). *Enterobacter asburiae* isolated from water courses in US contained plasmid having bla$_{IMI-2}$ genes producing IMI-2 carbapenemase enzymes making this strain resistant to imipenem (Rotova et al., 2017). *Enterobacter cloacae* shows resistance against ampicillin, cephalothin, cefoxitin and amoxicillin-clavulanic acid naturally and now have developed resistance against imipenem and β-lactams especially third generation cephalosporins except cefepime by overproduction of chromosomal cephalosporinase combined with porin alterations (Sanders and Sanders, 1997). Resistance rate against imipenem was found to be 0.4% in *Enterobacter cloacae* (Lee et al., 2005). Jacoby (2009) reported that production of chromosomally encoded AmpC β-lactamase in *Enterobacter cloacae* made them resistant to cephamycin and first and second generation cephalosporins. *Enterobacter hormaechei* strain of ECC is naturally sensitive to many antibiotics like penicillin, aminoglycosides, cefotaxime, lactams, quinolones etc. Resistance against third generation cephalosporins have been acquired by few strains of *E. hormaechei* as they harbor genes for production of AmpC cephalosporinases and extended-spectrum β-lactamases. Few carbapenem-resistant and fluoroquinolones-resistant strains of *E. hormaechei* were also reported by Hoffmann et al. (2005c). Some clones of *E. kobei* have also acquired resistance against β-lactams and carbapenems in Northeast China. These clones carry $bla_{\text{CTX-M-12}}$ gene on its chromosome which is responsible for the production of ESBLs (Zhou et al., 2017). All strains of *E. ludwigii* are resistant to many antibiotics including cefoxitin, amoxicillin-clavulanic acid and ampicillin naturally but are susceptible to imipenem, gentamicins, ciprofloxacin and cotrimoxazole. Some clinical strains also harbor genes for hyperproduction of AmpC cephalosporinase (Flores-Carrero et al., 2016). According to Hartl et al. (2018) $bla_{\text{IMI-2}}$ carbapenemase-producing genes were isolated from *E. mori* causing human infections in Austria. The coexistence of different drug resistant determinants on plasmids renders the development of multi-drug resistance in different bacterial strains (Blair et al., 2015). Presence of multi-drug resistant determinants including $bla_{\text{IMP-26}}$, $bla_{\text{DHA-1}}$ and $bla_{\text{SHV-12}}$ β-lactam-resistant, *aph(6)-Id*, *aacA4* and *aac*(6′)-*IIc* aminoglycoside-resistant, *aac*(6′)-*Ib3* and *qnrB4* fluoroquinolone-resistant and *fosA5* fosfomycin-resistant genes on pIMP26 plasmid of *Enterobacter cloacae* was reported by Wang et al. (2019). Furthermore, spread of ECC in any region depends upon the geographical and seasonal variations.

5.1. β-Lactam Resistant ECC

Enterobacter cloacae Complex has become resistant to widely used antibiotics named as β-lactams. Resistance against β-lactams is due the production of β-lactamases in some species of ECC. *Enterobacter cloacae* is resistant to almost all types of β-lactams by production of all four molecular Ambler classes of β-lactamases i.e., class A (penicillinases), class B (metalloenzymes), class C (cephalosporinases) and class D (oxacillinases) while other members of ECC are resistant to some of these classes. Extended spectrum β-lactamases-producing *E. cloacae* were first reported by DeChamps et al. (1989).

Table 1. List of ECC strains showing resistance against antibiotics

ECC Strain		Resistant to Antibiotics	References
1.	*E.asburiae*	Cefalotin, Tetracycline and Ampicillin, Imipenem.	Brenner et al. (1986), Rotova et al. (2017).
2.	*E. cloacae*	Ampicillin, Cephalothin, Cefoxitin, Amoxicillin-clavulanic acid, Imipenem, first second and third generation Cephalosporins, Cephamycin, tigecycline, doxycyclin, erythromycin, gentamycin, streptomycin, sulphamethoxazole/ trimethoprim.	Sanders and Sanders (1997), Lee et al. (2005), Jacoby (2009), Jiang et al. (2019), Jamal et al. (2019).
3.	*E. hormaechei*	Penicillin, Aminoglycosides, Cefotaxime, Lactams, Quinolones, Cephalosporins, Carbapenems, Avibactam, Cefiderocol.	Hoffmann et al. (2005c), Davin-Regli et al. (2019), Shields et al. (2020)
4.	*E. kobei*	β-lactams, Carbapenems.	Zhou et al. (2017)
5.	*E.ludwigii*	Cefoxitin, Amoxicillin-clavulanic acid, Ampicillin, Cephalosporins, Carbapenems.	Khajuria et al. (2013), Flores-Carrero et al. (2016).
6.	*E. mori*	Carbapenem	Hartl et al. (2018).

There are different genes present on chromosomes or plasmids that trigger the production of different β-lactamases for resistance against different classes of β-lactams. Naas et al. (2012) reported clinical isolates of *E. cloacae* harboring $bla_{\text{IMI-1}}$ gene causing resistance against β-lactams from France. *Enterobacter hormaechei* contains *ampC* and $bla_{\text{ACT-1}}$ genes responsible for the production of AmpC β-lactamases in it (Roh et al., 2012). The depression of AmpC β-lactamases is very frequent among clinical isolates of ECC producing resistance against ureido-penicillins, carboxy-penicillins and third generation cephalosporins thus these are not inhibited by clavulanate (a β-lactamase inhibitor). This AmpC-type β-lactamases resistance against β-lactams is shown by 50% clinical strains of *E. cloacae* as it is also responsible for the production of ESBLs. Yee-Huang et al. (2018) reported the presence of plasmid-mediated β-Lactamase genes in *E. cloacae* responsible for production of AmpC-type β-Lactamase from Southern Taiwan. Ku et al. (2018) have reported high prevalence rate of multiple plasmid-mediated *ampC* and ESBL-producing genes such as *bla*$_{CMH-1}$ and *bla*$_{MIR-6}$ genes, in *E. cloacae* in Southern Taiwan causing complexity in resistance mechanism and contribution in evolution also. Su et al. (2010) screened ESBL-producing isolates of *E. cloacae* in several hospitals.

Enterobacter cloacae is resistant to cefotaxime, ceftazidime and ceftriaxone. The resistance to these third generation cephalosporins is of major concern as these drugs are vital in the treatment of many ECC infections. The resistance against these drugs is due to production of ESBLs, AmpC-type β-lactamases and cephalosporinases. AmpC-type β-lactamases provides resistance against almost all types of β-lactams and cephamycins, a fourth generation cephalosporin (Mezzatesta et al., 2012). Shields et al. (2020) reported some strains of *E. hormaechei* isolated from two patients in America showing resistance against ceftazidime-avibactam and cefiderocol. AmpC β-lactamases can be inhibited by boronic acid, cloxacillin and fourth generation of cephalosporins (George, 2009).

Extended spectrum β-lactamase-producing ECC have ability to hydrolyze the fourth generation of cephalosporin (cefepime) and aztreonam (Szabo et al. 2005). Induction of ESBLs in ECC is the risk for modern medicine as it limits the choice of drugs for treating bacterial infection. Extended spectrum β-lactamase-producing genes are mostly present on plasmids while *ampC* genes are present on both plasmids and chromosomes. Treatment of infections caused by ECC having plasmid mediated ESBL or *ampC* genes is very complicated as these genes encode resistant traits for many other antibiotics (Mezzatesta et al., 2012 and Esteban-Cuesta et al., 2019). The methods for the detection of ESBLs including double-disk synergy test (DDST) and Vitek ESBLs detection test depend on the production of chromosomal enzymes and are expected to be very difficult. In case of overproduction of Bush group 1 β-lactamases, the detection methods for ESBLs is very inadequate (Tzelepi et al., 2000). Ho et al. (2005) reported that presence of *bla*$_{TEM}$, *bla*$_{SHV}$, *bla*$_{CTX-M}$ genes in *E. cloacae* causes production of ESBLs in Hong Kong. Genes encoding MBLs in ECC are present on plasmids and transposons and move between bacteria through horizontal transfer of these mobile genetic elements (Villa et al., 2014). IMP-type MBLs inactivates almost all β-lactams except monobactams (Bush, 2013). Clinical isolates of *E. cloacae* producing IMP-type MBLs has been reported by Dai et al. (2013) from Chongqing, China. Wang et al. (2017) reported two IMP-26-producing isolates of *E. cloacae* from bloodstream in Shanghai. Some *E. cloacae* strains showed resistance against MBLs by producing VIM-1-type MBLs (Falcone et al., 2009). Production of ESBLs can be inhibited by clavulanic acid (Choi et al., 2007).

5.1.1. Carbapenem Resistant ECC

Carbapenems are the most extensively used drug for nosocomial infections. The carbapenem-resistance in ECC is increasing over the period of time and is the major problem in the field of medicine as carbapenems are used as last level treatment of ECC infections when other antibiotics fail to treat it (Paterson and Bonomo, 2005). *Enterobacter cloacae* Complex has become second most prevalent carbapenemase producing bacteria worldwide (Alves et al., 2018). One of the reasons for the emergence of carbapenem resistance in ECC is heavy exposure to various antibiotics (Cai et al., 2019). Resistance against carbapenem is developed through multiple mechanisms like production of carbapenemases, structural changes leading to membrane impermeability, increased efflux of antibiotics and overproduction of β-lactamase or MBLs in ECC. *Enterobacter cloacae* Complex acquires drug resistant genes through plasmid-mediated transformations and horizontal transfer of plasmids spread carbapenemase encoding genes among different bacterial strains including clonal expansion spread of the structurally associated resistance traits (Zhao et al., 2020). some carbapenemase encoding genes in ST93 clusters of ECC spread in Vietnam, Thailand, Spain, Australia, China, Romania, Belgium and US are bla_{VIM-1}, bla_{IMP-14}, bla_{OXA-48}, bla_{NDM-1}, bla_{IMP-8} and bla_{KPC-2} (Peirano et al., 2014). *Enterobacter cloacae* and *E. hormaechei* harbouring bla_{VEB}, $bla_{GES/IBC}$, bla_{KPC} and bla_{FRI} genes showing resistance against carbapenems were reported by Kubota et al. (2018) in Tokyo, Japan. Boyd et al. (2017) reported the presence of bla_{NMC-A} and bla_{IMI} genes on plasmid or integrative elements of chromosome in ECC producing carbapenemases and Khajuria et al. (2013) for the first time reported the presence of carbapenem resistant *bla*$_{\text{NDM-1}}$ and *bla*$_{\text{OXA-48}}$ genes in bone infection causing *E. ludwigii.* Emergence of *bla*$_{\text{NMD}}$ and *bla*$_{\text{IMP}}$ genes in *E. cloacae* and *E. hormaechei* was reported by Bogaerts et al. (2011) in Belgium. NDM-producing ECC strains are very hazardous because of lack of effective antibiotics for their treatment and due to high prevalence of their unrecognized asymptomatic carriers (Rolain et al., 2010). Presence of multi-drug resistance genes, *bla*$_{\text{KPC-2}}$ and *bla*$_{\text{CTX-M}}$, in *E. hormaechei* and *E. asburiae* were reported by Andrade et al. (2018). In China, *Klebsiella pneumoniae* carbapenemase (KPC)-producing carbapenem resistant *E. cloacae* were also reported in 2010 from Shangai. KPC is a plasmid-mediated enzyme that hydrolyzes the carbapenems thus resist the carbapenems and that KPC enzyme along with *Arm*A methylase produces resistance against aminoglycoside in ECC that are already carbapenem resistant (Wu et al., 2010). KPC-producing clones ST78 and ST171 of *E. cloacae* were showing high-risk carbapenem resistance (Rees et al., 2018 and Gomez-Simmonds et al. 2018). Hamprecht et al. (2013) found that in Germany some isolates of *E. cloacae* showed resistance against carbapenems by production of GIM-1-type carbapenemase. Recent survey declares Balkan countries as reservoir of *bla*$_{\text{NDM-1}}$, a carbapenemase.. A variety of carbapenemase-producing genes are associated with high clonal diversity causing carbapenem resistance in ECC. Clones and clades containing carbapenemase genes and specific genetic elements were found to be spreading locally according to the study made by Peirano et al., 2014. Hyperproduction of carbapenem-hydrolyzing β-lactamase, overproduction of ESBLs along with decreased outer membrane permeability of drug and hyperproduction of AmpC-type cephalosporinase are other factor for inducing carbapenem resistance in ECC (Tian et al., 2020). Higher in-hospital mortality rate in Vietnam is found to be associated with carbapenemase-producing *E. cloacae* (Miyoshi-Akiyama et al., 2020).

5.2. Fluoroquinolones Resistant ECC

Quinolones are a class of antibiotics used in clinical therapy of infections caused by ECC and some members of ECC have become resistant to quinolones including fluoroquinolones. Mode of action of quinolones is to react with topoisomerases and inhibit replication of DNA leading to bacterial death (Drlica et al., 2008). Resistance against fluoroquinolones in ECC developed through different mechanisms. One of these mechanisms is by modification in the permeability of chromosomally mediated efflux pumps named as AcrAB and MexAB (Ruiz, 2003) or plasmid-mediated efflux pump *qepA* (Cano et al., 2009), other mechanisms are; genetic mutations in genes of DNA gyrase *gyrA*, *gyrB* leading to reduced binding of quinolones (Aldred et al., 2014) and mutations in genes of topoisomerase Ⅳ such as *qnr*, *aac*(6′)-Ib-cr and *qepA* genes (Ma et al., 2009). Plasmid-mediated quinolone resistance is determined by *qnrA, qnrB, qepA, qnrS, qnrC, qnrD* and *aac(6)-Ib* genes. *gyrA* and *parC* are the quinolone resistance determining regions in ECC (Zhao et al., 2020 and Gibson et al., 2010). Presence of plasmid-mediated *qepA* and *qnrS1* genes causing fluoroquinolone resistance in *Enterobacter aerogenes* along with β-lactamase-producing $bla_{\text{LAP-1}}$ gene was reported by Park et al. (2009). Wu et al. (2007) carried out a research on clinical isolates of *E. cloacae* in Taiwan and confirmed the presence of plasmid-mediated quinolone resistant genes, *qnrA*, *qnrB*, and *qnrS* in them. Xia et al. (2013) reported presence of *qnr* in multi-drug resistant *Enterobacter* spp. from Jinan, China. Mutations in *parC* and *gyrA* genes were detected in few isolates of ECC including *E. cloacae* showing resistance against quinolones (Deguchi et al., 1997).

5.3. Aminoglycosides Resistant ECC

Plasmid encoded aminoglycoside modifying enzymes including acetyltransferases, adenylyltransferases and phosphotransferases along with some transposable elements produce aminoglycosides resistance in

ECC. Resistance against amikacin and gentamicin was observed in of *E. cloacae* due to presence of *aac*(6′)-Ib gene (Kim et al., 2009). Resistance phenotypes are mostly acquired through either dispersion of transposons or exchange of plasmids. About 0-34% *E. cloacae* are resistant to amikacin and 0-51% are resistant to gentamicin (Sanders and Sanders, 1997). According to a research by Huang et al. (2012) in China, 77% of aminoglycoside resistance in ECC is plasmid-mediated. In Iran, some strains of *E. cloacae* were found to have *armA* and *rmtB* genes responsible for production of methylase making them resistant to aminoglycosides (Amin et al., 2019).

5.4. Tigecycline Resistant ECC

Tigecycline is regarded as last resort treatment to multi-drug resistant ECC by USFDA and is recommended for some serious and complicated infections (Sader et al., 2016). Majority of ECC are susceptible to this drug as it evades common mechanism of resistance showed by gram-negative bacteria. However, some studies showed that some clinical strains of ECC are showing certain resistance against tigecycline because of *RamA*-mediated overexpression of AcrAB efflux pump (Keeney et al., 2007). Many isolates of ECC belonging to Europe, North America, Latin America and Asia-Pacific rim showed increased resistance against minocycline (Xavier and Dowzicky, 2012). According to research made by Daurel et al., 2009, *E. hormaechei* isolated from liver transplanted patients induce resistance against tigecycline after tigecycline based therapy. *Enterobacter cloacae* Complex can also acquire tigecycline resistance traits after exposure to aminoglycosides, penicillin and fluoroquinolones (Jiang et al., 2019).

5.5. Colistin Resistant ECC

Usage of colistin, a polymyxin, was made limited in 1980s due to its toxicity but now it is being used in clinical therapy against several multi-drug resistant ECC. Polymyxins are reintroduced in clinical practices in spite of its increased risk of renal dysfunction development (Alves et al., 2018). Dose optimization of colistin and proper monitoring of patients is kept in consideration for treating infections caused by multi-drug resistant ECC (Falagas and Kasiakou, 2006). Some members of ECC have acquired resistance against colistin through changing the negatively charged surface of lipopolysaccharides by regulation of *pmrA* and *phoP* genes and plasmid-mediated colistin resistant *mcr-1* gene (Baron et al., 2017). Zeng et al. (2016) reported *E. cloacae* harbouring plasmid-mediated *mcr-1* genes. Presence of colistin-resistant *mcr*-1 gene and its transfer between bacterial strains horizontally was reported by Zheng et al. (2016). The effectiveness of these drugs can be increased by using combined therapy of different antibiotics. Due to development of multi-drug resistance in ECC, treatment is done by combination of different drug by determining the specific amount of each drug depending upon the patient and severity of disease (Lee and Doi, 2014). Due to toxic effects of polymyxins, the carbapenem resistant isolates of ECC are treated with combination of polymyxin with meropenem or tigecycline to reduce the dose of polymyxin if used alone (Alves et al., 2018).

6. Global Challenges

The world wide spread of ECC showing multi-drug resistance is the major challenge for the modern medicine and a global threat for the health of humans and domestic animals. According to WHO, 2018, new resistant traits and mechanisms developed by microbes is the biggest hurdle for the effective treatment of many infections. Emergence of resistance against β-lactams, carbapenems, aminoglycosides, fluoroquinolones, polymyxins, cefotaxims, cephalosporins and many other drugs is becoming major problem for healthcare workers and life risk for patients suffering from these microbes. Khan et al. (2017) described it as treat to control the infections in hospitals because of continual evolution of resistant markers under selection pressure. Rice (2008) referred ESBL and carbapenemase-producing Enterobacteriaceae as "ESKAPEE" pathogens because of their ability to escape the effects of antibiotics. Multi-drug resistant ECC (producing ESBLs and β-lactamases) cause different nosocomial infections that are so difficult to treat, even some strains have high epidemic potential. The treatment cost for such infections, being complicated and limited options of therapy, is increased as compare to infections caused by non-resistant microbes (Desvars-Larrive et al., 2019). In Netherland, there was an outbreak of infection caused by *E. hormaechei* strain that carry many drug resistant genes viz $bla_{CTX\text{-}M\text{-}9}$, *aadB*, *qurA1* and *tetA* on their conjugative plasmid. The alarming situation is that the ESBL-producing bacteria are now resistant to fluoroquinolones making the treatment of infections more challenging. According to Wu et al. (2010), the enzymes causing carbapenem resistance in ECC are also responsible for development of aminoglycoside resistance when combined with some other factors. Infections caused by carbapenem-resistant ECC results in prolonged hospitalization, expensive treatment and high death rate in patients (Kelly et al., 2017). The spread of carbapenem resistant pathogens in China was through tourist travelling. The resistant genes were transferred from one bacterial strain to others through horizontal transmission of plasmid. The spread of colistin-resistance through *mcr-1* gene can prove dangerous for whole world (Zheng et al., 2016). To prevent the outbreak of multi-drug resistant infections it is necessary to detect the emergence of resistance in microbes at early stages. Due to development of multi-drug resistance traits in ECC, there is emergence and spread of many well-adapted epidemic resistant clones during hospital outbreaks (Davin-Regli et al., 2019).

7. Strategies to Cope With the Challenges

WHO (2018) has established a ranking of medically important antibiotics for treatment of human infections caused by ECC. This ranking is constantly being revised with time. The sources for transmission of infection in humans need strict monitoring to stop the spread of infectious

ECC. The reservoirs like urban wildlife, seafood etc. that introduce the antimicrobial resistance genes in non-resistant strains need checks such as screening and monitoring of imported food items especially sea foods before consumption, to avoid the access of resistant genes to human communities through local bacterial populations (Brouwer et al., 2019). Reasonable antibiotic administration and continuous surveillance is needed to control the multi-drug resistance in high-risk patients (Jiang et al., 2019). A systematic approach to optimize the use of broad-spectrum drugs is needed in public health-care centers (Fishman et al., 2012). Zowawi et al. (2015) described strategies to cope with these challenges such as introduction of new antibiotics, developing improved diagnostic techniques for early detection of resistance, improving adherence for infection prevention or developing some non-antibiotic treatment procedures. Many pharmaceutical companies are not interested in new antibiotic development because of less profit so new economic model should be considered (Zheng et al., 2018). Phage therapy should be considered to treat infections caused by biofilm forming multi-drug resistant ECC. Phages are host specific, making phage therapy a specific, harmless and efficient approach to treat multi-drug resistant bacteria (Sieradzki et al., 2003). A single phage that can kill all the known strains of ECC has not been discovered yet, therefore isolation and characterization of phages against multi-drug resistance and biofilm forming bacteria is very important (Watnick and Kolter, 2000).

CONCLUSION

Enterobacter cloacae complex is responsible for not only infections worldwide but also development of resistance against many antibiotics, limiting the treatment strategies of ECC infections, even some strains of ECC have also developed resistance against drugs used as last resort of treatment creating a distressing situation. To combat multi-drug resistance in ECC, a major threat for global public health, continuous focused surveillance is needed in health related research institutes, hospitals and at community level to check or slow down the mechanisms of resistance development in ECC and development of new potent drugs against the bacterial group.

References

Aldred, K. J., Kerns, R. J. and Osheroff, N. (2014). Mechanism of quinolone action and resistance. *Biochemistry*, 53(10): 1565-1574.

Al-Tawfiq, J. A., Laxminarayan, R. and Mendelson, M. (2017). How should we respond to the emergence of plasmid-mediated colistin resistance in humans and animals? *International Journal of Infectious Diseases*, 54: 77-84.

Alves, P. H., Boff, R. T., Barth, A. L. and Martins, A. F. (2019). Synergy of polymyxin B, tigecycline and meropenem against carbapenem-resistant *Enterobacter cloacae* complex isolates. *Diagnostic Microbiology and Infectious Disease*, 94(1): 81-85.

Amin, M., Mehdipour, G. and Navidifar, T. (2019). High distribution of 16S rRNA methylase genes rmtB and armA among *Enterobacter cloacae* strains isolated from an Ahvaz teaching hospital, Iran." *Acta Microbiologica et Immunologica Hungarica*, 66(3): 337-348.

Andrade, L. N., Siqueira, T. E. S., Martinez, R. and Darini, A. L. C. (2018). Multidrug-resistant CTX-M-(15, 9, 2)-and KPC-2-producing *Enterobacter hormaechei* and *Enterobacter asburiae* isolates possessed a set of acquired heavy metal tolerance genes including a chromosomal *sil* operon (for acquired silver resistance). *Frontiers in Microbiology*, 9: 539.

Baron, S., Bardet, L., Dubourg, G., Fichaux, M. and Rolain J. M. (2017). *mcr-1* plasmid-mediated colistin resistance gene detection in an *Enterobacter cloacae* clinical isolate in France. *Journal of Global Antimicrobial Resistance*, 10: 35-36.

Bertrand, X. and Dowzicky, M. J. (2012). Antimicrobial susceptibility among gram-negative isolates collected from intensive care units in North America, Europe, the Asia-Pacific Rim, Latin America, the

Middle East, and Africa between 2004 and 2009 as part of the Tigecycline Evaluation and Surveillance Trial. *Clinical Therapeutics*, 34(1): 124-137.

Blair, J. M. A., Webber, M. A., Baylay, A. J., Ogbolu, D. O. and Piddock, L. J. V. (2015). Molecular mechanisms of antibiotic resistance. *Nature Reviews Microbiology*, 13(1): 42-51.

Bogaerts, P., Bouchahrouf, W., de Castro, R. R., Deplano, A., Berhin, C., Piérard, D., Denis, O. and Glupczynski, Y. (2011). Emergence of NDM-1-producing Enterobacteriaceae in Belgium. *Antimicrobial Agents and Chemotherapy*, 55(6): 3036-3038.

Boyd, D. A., Mataseje, L. F., Davidson, R., Delport, J. A., Fuller, J., Hoang, L., Lefebvre, B., Levett, P. N., Roscoe, D. L., Willey, B. M. and Mulvey, M. R. (2017). *Enterobacter cloacae* complex isolates harboring $bla_{\text{NMC-A}}$ or bla_{IMI}-type class A carbapenemase genes on novel chromosomal integrative elements and plasmids. *Antimicrobial Agents and Chemotherapy*, 61(5): e02578-16.

Brenner, D. J., McWorther, A. C., Kai, A., Steigerwalt, A. G., Farmer, J. J. (1986). *Enterobacter asburiae* sp. nov., a new species found in clinical specimens, and reassignment of *Erwinia dissolvens* and *Erwinia nimipressuralis* to the genus *Enterobacter* as *Enterobacter dissolvens* comb. nov. and *Enterobacter nimipressuralis* comb. nov. *Journal of Clinical Microbiology*, 23(6): 1114-1120.

Brkic, S., Bozic, D., Stojanovic, N., Vitorovic, T., Topalov, D., Jovanovic, M., Stepanovic, M. and Cirkovic, I. (2019). Antimicrobial Susceptibility and Molecular Characterization of Carbapenemase-Producing *Enterobacter* spp. Community Isolates in Belgrade, Serbia. *Microbial Drug Resistance*, 26(4): 378-384.

Brouwer, M. S. M., Tehrani, K. H. M. E., Rapallini, M., Geurts, Y., Kant, A., Harders, F., Mashayekhi, V., Martin, N. I., Bossers, A., Mevius, D. J., Wit, B. and Veldmana, K. T. (2019). Novel carbapenemases FLC-1 and IMI-2 encoded by an *Enterobacter cloacae* complex isolated from food products. *Antimicrobial Agents and Chemotherapy*, 63(6): e02338-18.

Bush, K. (2013). Carbapenemases: partners in crime. *Journal of Global Antimicrobial Resistance*, 1(1): 7-16.

Cai, Y., Chen, C., Zhao, M., Yu, X., Lan, K., Liao, K., Guo, P., Zhang, W., Ma, X., He, Y., Zeng, J., Chen, L., Jia, W., Tang, Y. W. and Huang, B. (2019). High Prevalence of Metallo-β-Lactamase-Producing *Enterobacter cloacae* from Three Tertiary Hospitals in China." *Frontiers in Microbiology*, 10: 1610.

Cano, M. E., Rodríguez-Martínez, J. M., Agüero, J., Pascual, A., Calvo, J., García-Lobo, J. M., Velasco, C., Francia, M. V. and Martínez-Martínez, L. (2009). Detection of plasmid-mediated quinolone resistance genes in clinical isolates of *Enterobacter* spp. in Spain. *Journal of Clinical Microbiology*, 47(7): 2033-2039.

Carter, M. Q., Pham, A., Huynh, S. and He, X. (2017). Complete genome sequence of a Shiga toxin-producing *Enterobacter cloacae* clinical isolate. *Genome Announcements*, 5(37): e00883-17.

Choi, S. H., Lee, J. E., Park, S. J., Kim, M. N., Choo, E. J., Kwak, Y. G., Jeong, J. Y., Woo, J. H., Kim, N. J. and Kim, Y. S. (2007). Prevalence, microbiology, and clinical characteristics of extended-spectrum β-lactamase-producing *Enterobacter* spp., *Serratia marcescens, Citrobacter freundii*, and *Morganella morganii* in Korea." *European Journal of Clinical Microbiology & Infectious Diseases*, 26(8): 557-561.

Curcio, D. (2014). Multidrug-resistant Gram-negative bacterial infections: are you ready for the challenge? *Current Clinical Pharmacology*, 9(1): 27-38.

Dai, W., Sun, S., Yang, P., Huang, S., Zhang, X. and Zhang, L. (2013). Characterization of carbapenemases, extended spectrum β-lactamases and molecular epidemiology of carbapenem-non-susceptible *Enterobacter cloacae* in a Chinese hospital in Chongqing. *Infection, Genetics and Evolution*, 14: 1-7.

Daurel, C., Fiant, A. L., Bremont, S., Courvalin, P. and Leclercq, R. (2009). Emergence of an *Enterobacter hormaechei* strain with reduced susceptibility to tigecycline under tigecycline therapy. *Antimicrobial Agents and Chemotherapy*, 53(11): 4953-4954.

Davin-Regli, A., Lavigne, J. P. and Pages, J. M. (2019). *Enterobacter* spp.: update on taxonomy, clinical aspects, and emerging antimicrobial resistance. *Clinical Microbiology Reviews*, 32(4): e00002-19.

Davin-Regli, Anne. "*Enterobacter aerogenes* and *Enterobacter cloacae*; versatile bacterial pathogens confronting antibiotic treatment." *Frontiers in microbiology* 6 (2015): 392.

DeChamps, C., Sauvant, M. P., Chanal, C., Sirot, D., Gazuy, N., Malhuret, R., Baguet, J. C., Sirot, J. (1989). Prospective survey of colonization and infection caused by expanded-spectrum-beta-lactamase-producing members of the family Enterobacteriaceae in an intensive care unit. *Journal of Clinical Microbiology*, 27(12): 2887-2890.

Deguchi, T., Yasuda, M., Nakano, M., Ozeki, S., Kanematsu, E., Nishino, Y., Ishihara, S. and Kawada, Y. (1997). Detection of mutations in the *gyrA* and *parC* genes in quinolone-resistant clinical isolates of *Enterobacter cloacae*. *The Journal of Antimicrobial Chemotherapy*, 40(4): 543-549.

Desvars-Larrive, A., Ruppitsch, W., Lepuschitz, S., Szostak, M. P., Spergser, J., Feßler, A. T., Schwarz, S., Monecke, S., Ehricht, R., Walzer, C. and Loncaric, I. (2019). Urban brown rats (*Rattus norvegicus*) as possible source of multidrug-resistant Enterobacteriaceae and meticillin-resistant *Staphylococcus* spp., Vienna, Austria, 2016 and 2017. *Eurosurveillance*, 24(32): 1900149.

Drlica, K., Malik M., Kerns R. J., and Zhao, X. (2008). Quinolone-mediated bacterial death. *Antimicrobial Agents and Chemotherapy*, 52(2): 385-392.

Esteban-Cuesta, I., Dorn-In, S., Drees, N., Holzel, C., Gottschalk, C., Gareis, M. and Schwaiger, K. (2019). Antimicrobial resistance of *Enterobacter cloacae* complex isolates from the surface of muskmelons. *International Journal of Food Microbiology*, 301: 19-26.

Falagas, M. E., Kasiakou, S. K. and Saravolatz, L. D. (2005). Colistin: the revival of polymyxins for the management of multidrug-resistant gram-negative bacterial infections. *Clinical Infectious Diseases*, 40(9): 1333-1341.

Falcone, M., Mezzatesta, M. L., Perilli, M., Forcella, C., Giordano, A., Cafiso, V., Amicosante, G., Stefani, S. and Venditti, M. (2009). Infections with VIM-1 metallo-β-lactamase-producing *Enterobacter cloacae* and their correlation with clinical outcome. *Journal of Clinical Microbiology*, 47(11): 3514-3519.

Fernández, L. and Hancock, R. E. W. (2012). Adaptive and mutational resistance: role of porins and efflux pumps in drug resistance. *Clinical Microbiology Reviews*, 25(4): 661-681.

Fernandez-Baca, V., Ballesteros, F., Hervas, J. A., Villalon, P., Domínguez, M. A., Benedí, V. J. and Albertí, S. (2001). Molecular epidemiological typing of *Enterobacter cloacae* isolates from a neonatal intensive care unit: three-year prospective study. *Journal of Hospital Infection*, 49(3): 173-182.

Fishman, N., Society for Healthcare Epidemiology of America, Infectious Diseases Society of America and Pediatric Infectious Diseases Society. (2012). Policy statement on antimicrobial stewardship by the society for healthcare epidemiology of America (SHEA), the infectious diseases society of America (IDSA), and the pediatric infectious diseases society (PIDS). *Infection Control & Hospital Epidemiology*, 33(4): 322-327.

Flores-Carrero, A., Labrador, I., Paniz-Mondolfi, A., Peaper, D. R., Towle, D. and Araque, M. (2016). Nosocomial outbreak of extended-spectrum β-lactamase-producing *Enterobacter ludwigii* co-harbouring CTX-M-8, SHV-12 and TEM-15 in a neonatal intensive care unit in Venezuela. *Journal of Global Antimicrobial Resistance*, 7: 114-118.

Gibson, J. S., Cobbold, R. N., Heisig, P., Sidjabat, H. E., Kyaw-Tanner, M. T. and Trott, D. J. (2010). Identification of Qnr and AAC (6′)-1b-cr plasmid-mediated fluoroquinolone resistance determinants in multidrug-resistant *Enterobacter* spp. isolated from extraintestinal infections in companion animals. *Veterinary Microbiology*, 143(2): 329-336.

Gomez-Simmonds, A., Annavajhala, M. K., Wang, Z., Macesic, N., Hu, Y., Giddins, M. J., O'Malley, A., et al. (2018). Genomic and geographic context for the evolution of high-risk carbapenem-resistant

Enterobacter cloacae complex clones ST171 and ST78. *American Society of Microbiology*, 9(3): e00542-18.

Guerin, F., Isnard, C., Cattoir, V. and Giard, J. C. (2015). Complex regulation pathways of AmpC-mediated β-lactam resistance in *Enterobacter cloacae* complex. *Antimicrobial Agents and Chemotherapy*, 59(12): 7753-7761.

Hamprecht, A., Poirel, L., Gottig, S., Seifert, H., Kaase, M. and Nordmann, P. (2013). Detection of the carbapenemase GIM-1 in *Enterobacter cloacae* in Germany. *Journal of Antimicrobial Chemotherapy*, 68(3): 558-561.

Harbarth, S., Sudre, P., Dharan, S., Cadenas, M. and Pittet, D. (1999). Outbreak of *Enterobacter cloacae* related to understaffing, overcrowding, and poor hygiene practices. *Infection Control & Hospital Epidemiology*, 20(9): 598-603.

Hartl, R., Kerschner, H., Gattringer, R., Lepuschitz, S., Allerberger, F., Sorschag, S., Ruppitsch W. and Apfalter, P. (2019). Whole-genome analysis of a human *Enterobacter mori* isolate carrying a $bla_{\mathrm{IMI-2}}$ carbapenemase in Austria. *Microbial Drug Resistance*, 25(1): 94-96.

Ho, P. L., Shek, R. H. L., Chow, K. H., Duan, R. S., Mak, G. C., Lai, E. L., Yam, W. C., Tsang, K. W. and Lai, W. M. (2005). Detection and characterization of extended-spectrum β-lactamases among bloodstream isolates of *Enterobacter* spp. in Hong Kong, 2000–2002. *Journal of Antimicrobial Chemotherapy*, 55(3): 326-332.

Hoffmann, H., Schmoldt, S., Trülzsch, K., Stumpf, A., Bengsch, S., Blankenstein, T., Heesemann, J. and Roggenkamp, A. (2005a). Nosocomial urosepsis caused by *Enterobacter kobei* with aberrant phenotype. *Diagnostic Microbiology and Infectious Disease*, 53(2): 143-147.

Hoffmann, H., Stindl, S., Ludwig, W., Stumpf, A., Mehlen, A., Monget, D., Pierard, D., Ziesing, S., Heesemann, J., Roggenkamp, A. and Schleifer, K. H. (2005c). *Enterobacter hormaechei* subsp. *oharae* subsp. nov., *E. hormaechei* subsp. *hormaechei* comb. nov., and *E. hormaechei* subsp. *steigerwaltii* subsp. nov., three new subspecies of clinical importance. *Journal of Clinical Microbiology*, 43(7): 3297-3303.

Hoffmann, H., Stindl, S., Stumpf, A., Mehlen, A., Monget, D., Heesemann, J., Schleifer, K. H. and Roggenkamp, A. (2005b). Description of *Enterobacter ludwigii* sp. nov., a novel *Enterobacter* species of clinical relevance. *Systematic and Applied Microbiology*, 28(3): 206-212.

Huang, L., Feng, Y. and Zong, Z. (2019). Heterogeneous resistance to colistin in *Enterobacter cloacae* complex due to a new small transmembrane protein. *Journal of Antimicrobial Chemotherapy*, 74(9): 2551-2558.

Huang, S., Dai, W., Sun, S., Zhang, X. and Zhang, L. (2012). Prevalence of plasmid-mediated quinolone resistance and aminoglycoside resistance determinants among carbapeneme non-susceptible *Enterobacter cloacae*. *PloS one*, 7(10): e47636.

Humann, J. L., Wildung, M., Cheng, C. H., Lee, T., Stewart, J. E., Drew, J. C., Triplett, E. W., Main, D. and Schroeder, B. K. (2011). Complete genome of the onion pathogen *Enterobacter cloacae* EcWSU1. *Standards in Genomic Sciences*, 5(3): 279-286.

Izdebski, R., Baraniak, A., Herda, M., Fiett, J., Bonten, M. J. M., Carmeli, Y., Goossens, H., Hryniewicz, W., Brun-Buisson, C. and Gniadkowski, M. (2015). MLST reveals potentially high-risk international clones of *Enterobacter cloacae*. *Journal of Antimicrobial Chemotherapy*, 70(1): 48-56.

Jacoby, G. A. (2009). AmpC β-lactamases. *Clinical Microbiology Reviews*, 22(1): 161-182.

Jamal, M., Andleeb, S., Jalil, F., Imran, M., Nawaz, M. A., Hussain, T., Ali, M., Rahman, S. and Das, C. R. (2019). Isolation, characterization and efficacy of phage MJ2 against biofilm forming multi-drug resistant *Enterobacter cloacae*. *Folia Microbiologica*, 64(1): 101-111.

Jean, S. S. and Hsueh, P. R. (2011). High burden of antimicrobial resistance in Asia. *International Journal of Antimicrobial Agents*, 37(4): 291-295.

Jiang, Y., Jia, X. and Xia, Y. (2019). Risk factors with the development of infection with tigecycline-and carbapenem-resistant *Enterobacter cloacae*. *Infection and Drug Resistance*, 12: 667-674.

Keeney, D., Ruzin, A. and Bradford, P. A. (2007). RamA, a transcriptional regulator, and AcrAB, an RND-type efflux pump, are associated with decreased susceptibility to tigecycline in *Enterobacter cloacae*. *Microbial Drug Resistance*, 13(1): 1-6.

Kelly, A. M., Mathema, B. and Larson, E. L. (2017). Carbapenem-resistant Enterobacteriaceae in the community: a scoping review. *International Journal of Antimicrobial Agents*, 50(2): 127-134.

Khajuria, A., Praharaj, A. K., Grover, N. and Kumar, M. (2013). First report of an *Enterobacter ludwigii* isolate coharboring NDM-1 and OXA-48 carbapenemases. *Antimicrobial Agents and Chemotherapy*, 57(10): 5189-5190.

Khan, A. U., Maryam, L. and Zarrilli, R. (2017). Structure, genetics and worldwide spread of New Delhi metallo-β-lactamase (NDM): a threat to public health. *BMC Microbiology*, 17(1): 101.

Kim, D. M., Jang, S. J., Neupane, G. P., Jang, M. S., Kwon, S. H., Kim, S. W. and Kim, W. Y. (2010). *Enterobacter nimipressuralis* as a cause of pseudobacteremia. *BMC Infectious Diseases*, 10(1): 315.

Kim, S. M., Lee, H. W., Choi, Y. W., Kim, S. H., Lee, J. C., Lee, Y. C., Seol, S. Y., Cho, D. T. and Kim, J. (2012). Involvement of curli fimbriae in the biofilm formation of *Enterobacter cloacae*. *The Journal of Microbiology*, 50(1): 175-178.

Kim, S. Y., Park, Y. J., Yu, J. K., Kim, Y. S. and Han, K. (2009). Prevalence and characteristics of *aac(6')-Ib-cr* in AmpC-producing *Enterobacter cloacae*, *Citrobacter freundii*, and *Serratia marcescens*: a multicenter study from Korea. *Diagnostic Microbiology and Infectious Disease*, 63(3): 314-318.

Ku, Y. H., Lee, M. F., Chuang, Y. C. and Yu, W. L. (2019). Detection of plasmid-mediated β-Lactamase genes and emergence of a novel AmpC (CMH-1) in *Enterobacter cloacae* at a medical Center in Southern Taiwan. *Journal of Clinical Medicine*, 8(1): 8.

Kubota, H., Uwamino, Y., Matsui, M., Sekizuka, T., Suzuki, Y., Okuno, R., Uchitani, Y., Ariyoshi, T., Aoki, W., Suzuki, S., Kuroda, M., Shinkai, T., Yokoyama, K., Sadamasu, K., Funakoshi, T., Murata, M., Hasegawa, N. and Iwata, S. (2018). FRI-4 carbapenemase-producing *Enterobacter cloacae* complex isolated in Tokyo, Japan. *Journal of Antimicrobial Chemotherapy*, 73(11): 2969-2972.

Lee, C. S. (2014). Therapy of infections due to carbapenem-resistant gram-negative pathogens. *Infection & Chemotherapy*, 46(3): 149-164.

Lee, H. K., Park, Y. J., Kim, J. Y., Chang, E., Cho, S. G., Chae, H. S. and Kang, C. S. (2005). Prevalence of decreased susceptibility to carbapenems among *Serratia marcescens*, *Enterobacter cloacae*, and *Citrobacter freundii* and investigation of carbapenemases. *Diagnostic Microbiology and Infectious Disease*, 52(4): 331-336.

Liu, C. P., Wang, N. Y., Lee, C. M., Weng, L. C., Tseng, H. K., Liu, C. W., Chiang, C. S. and Huang, F. Y. (2004). Nosocomial and community-acquired *Enterobacter cloacae* bloodstream infection: risk factors for and prevalence of SHV-12 in multiresistant isolates in a medical centre. *Journal of Hospital Infection*, 58(1): 63-77.

Liu, F., Yang, J., Xiao, Y., Li, L., Yang, F. and Jin, Q. (2016). Complete genome sequence of a clinical isolate of *Enterobacter asburiae*. *Genome announcements*, 4(3): e00523-16.

Ma, J., Zeng, Z., Chen, Z., Xu, X., Wang, X., Deng, Y., Lü, D., et al. High prevalence of plasmid-mediated quinolone resistance determinants *qnr*, *aac (6')-Ib-cr*, and *qepA* among ceftiofur-resistant Enterobacteriaceae isolates from companion and food-producing animals. *Antimicrobial Agents and Chemotherapy*, 53(2): 519-524.

Majewski, P., Wieczorek, P., Ojdana, D., Sieńko, A., Kowalczuk, O., Sacha, P., Nikliński, J. and Tryniszewska, E. (2016). Altered outer membrane transcriptome balance with AmpC overexpression in carbapenem-resistant *Enterobacter cloacae*. *Frontiers in Microbiology*, 7: 2054.

Mezzatesta, M. L., Gona1, F. and Stefani, S. (2012). *Enterobacter cloacae* complex: clinical impact and emerging antibiotic resistance. *Future Microbiology*, 7(7): 887-902.

Miyoshi-Akiyama, T., Ohmagari, N., Phuong, T. T., Huy, N. Q., Anh, N. Q., Thuy, P. T. P., Kirikae, T., Nhung, P. H. and Takemoto, N. (2020). Epidemiology of *Enterobacter cloacae* strains producing a carbapenemase or metallo-beta-lactamase in Vietnamese clinical settings in 2014–2017. *Journal of Medical Microbiology*, 69(4): 530-536.

Naas, T., Cattoen, C., Bernusset, S., Cuzon, G. and Nordmann, P. (2012). First identification of $bla_{\mathrm{IMI\text{-}1}}$ in an *Enterobacter cloacae* clinical isolate from France. *Antimicrobial Agents and Chemotherapy*, 56(3): 1664-1665.

O'hara, C. M., Steigerwalt, A. G., Hill, B. C., Farmer, J. J., Fanning, G. G. and Brenner, D. J. (1989). *Enterobacter hormaechei*, a new species of the family Enterobacteriaceae formerly known as enteric group 75. *Journal of Clinical Microbiology*, 27(9): 2046-2049.

Park, Y. J., Yu, J. K., Kim, S. I., Lee, K. and Arakawa, Y. (2009). Accumulation of plasmid-mediated fluoroquinolone resistance genes, *qepA* and *qnrS1*, in *Enterobacter aerogenes* co-producing RmtB and class A β-lactamase LAP-1. *Annals of Clinical & Laboratory Science*, 39(1): 55-59.

Paterson, D. L. and Bonomo, R. A. (2005). Extended-spectrum β-lactamases: a clinical update. *Clinical Microbiology Reviews*, 18(4): 657-686.

Peirano, G., Matsumura, Y., Adams, M. D., Bradford, P., Motyl, M., Chen, L., Kreiswirth, B. N. and Pitout, J. D. (2018). Genomic epidemiology of global carbapenemase-producing *Enterobacter* spp., 2008–2014. *Emerging Infectious Diseases*, 24(6): 1010.

Rees, C. A., Nasir, M., Smolinska, A., Lewis, A. E., Kane, K. R., Kossmann, S. E., Sezer, O., et al. (2018). Detection of high-risk carbapenem-resistant *Klebsiella pneumoniae* and *Enterobacter cloacae* isolates using volatile molecular profiles. *Scientific Reports*, 8(1): 1-13.

Ren, Y., Ren, Y., Zhou, Z., Guo, X., Li, Y., Feng, L. and Wang, L. (2010). Complete genome sequence of *Enterobacter cloacae* subsp. *cloacae* type strain ATCC 13047. *Journal of Bacteriology*, 192(9): 2463-2464.

Rice, L. B. (2008). Federal funding for the study of antimicrobial resistance in nosocomial pathogens: no ESKAPE. *The Journal of Infectious Diseases*, 197(8): 1079-1081.

Roh, K. H., Song, W., Chung, H. S., Lee, Y. S., Yum, J. H., Yi, H. N., Chun, J. S., Yong, D., Lee, K. and Chong., Y. (2012). Chromosomal cephalosporinase in *Enterobacter hormaechei* as an ancestor of ACT-1 plasmid-mediated AmpC β-lactamase. *Journal of Medical Microbiology*, 61(1): 94-100.

Rolain, J. M., Parola, P. and Cornaglia, G. (2010). New Delhi metallo-beta-lactamase (NDM-1): towards a new pandemia?. *Clinical Microbiology and Infection*, 16(12): 1699-1701.

Rotova, V., Papagiannitsis, C. C., Chudejova, K., Medvecky, M., Skalova, A., Adamkova, V. and Hrabak, J. (2017). First description of the emergence of *Enterobacter asburiae* producing IMI-2 carbapenemase in the Czech Republic. *Journal of Global Antimicrobial Resistance*, 11: 98-99.

Ruiz, J. (2003). Mechanisms of resistance to quinolones: target alterations, decreased accumulation and DNA gyrase protection. *Journal of Antimicrobial Chemotherapy*, 51(5): 1109-1117.

Sader, H. S., Castanheira, M., Farrell, D. J., Flamm, R. K., Mendes, R. E. and Jones, R. N. (2016). Tigecycline antimicrobial activity tested against clinical bacteria from Latin American medical centres: results from SENTRY Antimicrobial Surveillance Program (2011–2014). *International Journal of Antimicrobial Agents*, 48(2): 144-150.

Sanders, W. E. and Sanders, C. C. (1997). *Enterobacter* spp.: pathogens poised to flourish at the turn of the century. *Clinical Microbiology Reviews*, 10(2): 220-241.

Shields, R. K., Iovleva, A., Kline, E. G., Kawai, A. and McElheny, C. L. (2020). Clinical evolution of AmpC-mediated ceftazidime-avibactam and cefiderocol resistance in *Enterobacter cloacae* complex following exposure to cefepime. *Clinical Infectious Diseases*, https://doi.org/10.1093/cid/ciaa355.

Sieradzki, K., Leski, T., Dick, J., Borio, L. and Tomasz, A. (2003). Evolution of a vancomycin-intermediate *Staphylococcus aureus* strain in vivo: multiple changes in the antibiotic resistance phenotypes of a single lineage of methicillin-resistant *S. aureus* under the impact of antibiotics administered for chemotherapy. *Journal of Clinical Microbiology*, 41(4): 1687-1693.

Soi, V., Moore, C. L., Kumbar, L. and Yee, J. (2016). Prevention of catheter-related bloodstream infections in patients on hemodialysis: challenges and management strategies. *International Journal of Nephrology and Renovascular Disease*, 9: 95-103.

Su, P. A., Wu, L. T., Cheng, K. C., Ko, W. C., Chuang, Y. C. and Yu, W. L. (2010). Screening extended-spectrum β-lactamase production in *Enterobacter cloacae* and *Serratia marcescens* using antibiogram-based methods. *Journal of Microbiology, Immunology and Infection*, 43(1): 26-34.

Szabo, D., Melan, M. A., Hujer, A. M., Bonomo, R. A., Hujer, K. M., Bethel, C. R., Kristóf, K. and Paterson, D. L. (2005). Molecular analysis of the simultaneous production of two SHV-type extended-spectrum beta-lactamases in a clinical isolate of *Enterobacter cloacae* by using single-nucleotide polymorphism genotyping. *Antimicrobial Agents and Chemotherapy*, 49(11): 4716-4720.

Tian, X., Huang, C., Ye, X., Jiang, H., Zhang, R., Hu, X. and Xu, D. (2020). Carbapenem-Resistant *Enterobacter cloacae* Causing Nosocomial Infections in Southwestern China: Molecular Epidemiology, Risk Factors, and Predictors of Mortality. *Infection and Drug Resistance*, 13: 129-137.

Tzelepi, E., Giakkoupi, P., Sofianou, D., Loukova, V., Kemeroglou, A. and Tsakris, A. (2000). Detection of Extended-Spectrum β-Lactamases in Clinical Isolates of *Enterobacter cloacae* and *Enterobacter aerogenes*. *Journal of Clinical Microbiology*, 38(2): 542-546.

Van Boeckel, T. P., Gandra, S., Ashok, A., Caudron, Q., Grenfell, B. T., Levin, S. A. and Laxminarayan, R. (2014). Global antibiotic consumption 2000 to 2010: an analysis of national pharmaceutical sales data. *The Lancet Infectious Diseases*, 14(8): 742-750.

Villa, J., Viedma, E., Brañas, P., Orellana, M. A., Otero, J. R. and Chaves, F. (2014). Multiclonal spread of VIM-1-producing *Enterobacter cloacae* isolates associated with *In624* and *In488* integrons located in an *IncHI2* plasmid. *International Journal of Antimicrobial Agents*, 43(5): 451-455.

Wang, S., Xiao, S. Z., Gu, F. F., Tang, J., Guo, X. K., Ni, Y. X., Qu, J. M. and Han, L. Z. (2017). Antimicrobial susceptibility and molecular epidemiology of clinical *Enterobacter cloacae* bloodstream isolates in Shanghai, China. *PloS one*, 12(12): e0189713.

Wang, S., Zhou, K., Xiao, S., Xie, L., Gu, F., Li, X., Ni, Y., Sun, J. and Han, L. (2019). A Multidrug Resistance Plasmid pIMP26, Carrying $bla_{\text{IMP-26}}$, *fosA5*, $bla_{\text{DHA-1}}$, and *qnrB4* in *Enterobacter cloacae*. *Scientific Reports*, 9(1): 1-7.

Watnick, P. and Kolter, R. (2000). Biofilm, city of microbes. *Journal of Bacteriology*, 182(10): 2675-2679.

WHO. (2018). *Antimicrobial Resistance Fact Sheets*. World Health Organization Retrieved from. http://www.who.int/en/news-room/fact-sheets/detail/antimicrobialresistance/ (October 2018).

Wozniak, A., Villagra, N. A., Undabarrena, A., Gallardo, N., Keller, N., Moraga, M., Román, J. C., Mora, G. C. and García, P. (2012). Porin alterations present in non-carbapenemase-producing Enterobacteriaceae with high and intermediate levels of carbapenem resistance in Chile. *Journal of Medical Microbiology*, 61(9): 1270-1279.

Wu, J. J., Ko, W. C., Tsai, S. H. and Yan, J. J. (2007). Prevalence of plasmid-mediated quinolone resistance determinants *QnrA*, *QnrB*, and *QnrS* among clinical isolates of *Enterobacter cloacae* in a Taiwanese hospital. *Antimicrobial Agents and Chemotherapy*, 51(4): 1223-1227.

Wu, Q., Liu, Q., Han, L., Sun, J. and Ni, Y. (2010). Plasmid-mediated carbapenem-hydrolyzing enzyme KPC-2 and ArmA 16S rRNA methylase conferring high-level aminoglycoside resistance in carbapenem-resistant *Enterobacter cloacae* in China. *Diagnostic Microbiology and Infectious Disease*, 66(3): 326-328.

Xia, R., Ren, Y. and Xu, H. (2013). Identification of plasmid-mediated quinolone resistance *qnr* genes in multidrug-resistant Gram-negative bacteria from hospital wastewaters and receiving waters in the Jinan area, China." *Microbial Drug Resistance*, 19(6): 446-456.

Zeng, K. J., Doi, Y., Patil, S., Huang, X. and Tian, G. B. (2016). Emergence of the plasmid-mediated *mcr-1* gene in colistin-resistant *Enterobacter aerogenes* and *Enterobacter cloacae. Antimicrobial Agents and Chemotherapy*, 60(6): 3862-3863.

Zheng, B., Dong, H., Xu, H., Lv, J., Zhang, J., Jiang, X., Du, Y., Xiao, Y, and Li, L. (2016). Coexistence of MCR-1 and NDM-1 in clinical *Escherichia coli* isolates. *Clinical Infectious Diseases*, 63(10): 1393-1395.

Zheng, W., Sun, W., and Simeonov, A. (2018). Drug repurposing screens and synergistic drug-combinations for infectious diseases. *British Journal of Pharmacology*, 175(2): 181-191.

Zhou, K., Yu, W., Bonnet, R., Cattoir, V., Shen, P., Wang, B., Rossen, J. W. and Xiao, Y. (2017). Emergence of a novel *Enterobacter kobei* clone carrying chromosomal-encoded CTX-M-12 with diversified pathogenicity in northeast China. *New Microbes and New Infections*, 17: 7-10.

Zowawi, H. M., Harris, P. N. A., Roberts, M. J., Tambyah, P. A., Schembri, M. A., Pezzani, M. D., Williamson, D. A. and Paterson, D. L. (2015). The emerging threat of multidrug-resistant Gram-negative bacteria in urology. *Nature Reviews Urology*, 12(10): 570.

In: Recent Developments in *Enterobacter* … ISBN: 978-1-53618-615-4
Editor: Grégoire Roux

Chapter 5

RECENT TRENDS IN TAXONOMY OF GENUS *ENTEROBACTER*

Nazish Mazhar Ali, PhD and Iram Liaqat*, PhD
Department of Zoology, GC University,
Lahore, Pakistan

ABSTRACT

In recent times, the taxonomy of genus *Enterobacter* has gone through significant revision. The recent classification is based on DNA-DNA hybridization technique and 16S rRNA sequence analysis. Now this genus contains fifteen named species. Many members of this genus have already been transferred to other genera of Enterobacteriaceae family. These genera are; *Klebsiella*, *Serratia*, *Hafnia*, and *Pantoea*. This transfer has significant current genetic evidence in support. Transfer of *Enterobacter aerogenes* to genus *Klebsiella* and reclassification of a number of geno-species in the E. *cloacae* complex of organisms to different named species are good recent examples. On the other hand, some members of genus *Erwinia* have been transferred to genus *Enterobacter*. The *Enterobacter* species most commonly isolated from human sources are; *E. cloacae, E. aerogenes*, and to a lesser extent,

* Corresponding Author's Email:iramliaq@hotmail.com.

Pantoea (previously *Enterobacter*) *agglomerans* group, *Cronobacter* spp., *E. gergoviae,* and *E. asburiae*. *E. cloacae* is most common clinical isolate from human samples. Many newly named species (formerly called *E. cloacae* group) are difficult to differentiate only by morphological or biochemical test methods. So, advance molecular techniques are used for their classification in novel way. Matrix-assisted laser desorption/ionization–time of flight (MALDI-TOF) mass spectrometry is very helpful in identification of gram-negative bacilli and differentiation among *Enterobacter, Cronobacter,* and *Pantoea* species.

Keywords: *Enterobacter* spp., Enterobacteriaceae, phylogeny, molecular techniques, classification of genus *Enterobacter*

1. Genus *Enetrobacter* Taxonomy-Brief History

The genus *Enterobacter* is recognized as facultative anaerobic Gram-negative bacillus and belongs to family Enterobacteriaceae. This bacterium is 2 mm in length and is motile by use of peritrichous flagella. *Enterobacter* was first classified in 1960 but its taxonomy has gone through many changes during last five decades. *E. sakazakii* is an example as its name has been changed to a new genus, Cronobacter [1-2].

2. *Enterobacter* Species (Recent Nomenclature)

Till today, about 22 species have been included in the genus *Enterobacter.* These species include; *E. arachidis, E. asburiae, E. dissolvans, E. gergoviae, E. helveticus, E. hormaechei, E. kobei, E. ludwigii, E. mori, E. carcinogenus, E. cloacae, E. cowanii, E. nimipressuralis, E. oryzae, E. pulveris, E. pyrinus, E. radicincitans, E. soli, E. taylorae,* and *E. aerogenes, E. amnigenus, E. turicensi.* Seven of these species are assembled in the "*Enterobacter cloacae* complex group." This complex consists of *E. cloacae, E. nimipressuralis, E. asburiae, E. hormaechei, E. kobei, E. ludwigii, E. mori*. Nomenclature of this group is based on sharing of morphological and particularly genotypic characteristics. These characteristics are determined by DNA-DNA hybridizations of whole genome of these species. For example, *E. cloacae* has least 60% similarity of its genome with other members of this group [3, 4].

3. MAJOR HABITATS OF *ENTEROBACTER* SPECIES

The genus *Enterobacter* is found in a variety of environments *i.e.,* associated with different habitats. These bacteria can be isolated from soil, sewage and water and are considered as phytopathogens for many plant species. Few *Enterobacter* species are associated with metabolic engineering approaches and bioprocessing [5-6]. Moreover, *Enterobacter* spp. also act as natural commensals of animals and human especially as gut microbiota. Among these bacteria, only certain species or subspecies are associated with nosocomial infections and outbreaks [7-9]. *Enterobacter* species are one of members of "ESKAPE" group (*Enterococcus faecium, Staphylococcus aureus, Klebsiella pneumoniae, Acinetobacter baumannii, Pseudomonas aeruginosa,* and *Enterobacter* species). "ESKAPE" group members are considered as frequent cause of hospital acquired infections [10-18]. *E. aerogenes, E. cloacae,* and *E. hormaechei* represent the most frequently isolated species in hospital infections, particularly in immune-compromised individuals and also those admitted in an intensive care unit (ICU). Due to resistance of these species to antibacterial agents and also their behavior as opportunistic pathogens, are involved in several hospital outbreaks [19-26]. Antibiotic resistance, regulation of resistance genes and the clinical implications of these situations have been extensively studied [26-31].

4. Identification and Characterization of *Enterobacter* Species (Phylogeny)

The precise identification and characterization of *Enterobacter* species and subspecies is a big a challenge. In this regard, development of genome sequencing has resulted in modifications in phylogeny of this genus, particularly that of the *E. cloacae* complex [32, 34].

4.1. To/From *Enterobacter* (Modern Techniques as Tool for Re-Classification)

Due to use of modern molecular techniques, the genus has undergone modifications in classification, and several species have been transferred to and from this genus. First four species (*E. cowanii, E. arachidis, E. oryzae,* and *E. radicinintans*) have been reclassified to new genus *Kosakonia*. Another spcies, *E. intermedium* is reclassified to the genus *Kluyvera*. *E. sakazakii* is reclassified to the genus *Cronobacter* [2, 35, 36]. The current taxonomic position of *E. aerogenes* is still under discussion. The proposition of reclassification in the genus *Klebsiella* as *K. aerogenes, K. mobilis,* or *K. aeromobilis* because of its unsettled motility since 1971 [37]. The morphological differences between

E. aerogenes and the genus *Klebsiella* include motility, presence of ornithine decarboxylase and lack of urease activity in *E. aerogenes*. However, K. pneumoniae is closest species to *E. aerogenes* after results from full-genome sequence analysis [37].

4.2. Multilocus Sequence Analysis (MLSA) and Phylogeny of *Enterobacter*

Until now, scientific reports have occasionally used *K. aerogenes* and usually have used *E. aerogenes*. Multilocus sequence analysis (MLSA) of housekeeping genes in part, and 16S rRNA gene sequencing has recently allowed the characterization of new *Enterobacter* species [39-40]. The total genome sequences of the various *Enterobacter* sp. have cleared the path for reevaluation of phylogeny of this genus and better evaluation of the importance of those cases where some species were misidentified as other species by routine identification techniques, as was the case of *E. hormaechei* in the *E. cloacae* complex [41].

Table 1. Biochemical characteristics for the identification of species of the genus *Enterobacter*

Organism	Yellow pigment	LDC	ADH	URE	ESC	Fermentation								
						INO	SOR	SAC	MEL	RAF	AMG	DUL	ADO	ARL
E. amnigenus biogroup 1	–	–	–	–	+	–	–	+	+	+	V	–	–	–
E. amnigenus biogroup 2	–	–	V	–	+	–	+	–	+	–	+	–	–	–
E. cancerogenus	–	–	+	–	+	–	–	–	–	–	–	–	–	–
E. asburiae	–	–	(–)	+/–	+	–	+	+	V	(+)	+	–	–	–
E. cloacae subsp. *cloacae*	–	–	+	–	–	(–)	+	+	+	(+)	(+)	(–)	(–)	(–)
E. cloacae subsp. *dissolvens*	–	–	+	+w	+	V	+	+	+	+	+	–	–	–
E. hormaechei subsp. *hormaechei*	–	–	(+)	(+)	–	–	–	+	–	–	+	+	–	–
E. kobei	–	–	+	+	–	+	+	+	+	+	+	V	–	–
E. ludwigii	–	(–)	+	–	–	+	+	+	+	+	+	–	–	
E. nimipressuralis	–	–	–	–	+	–	+	–	+	–	+	–	–	
E. aerogenes	–	+	–	–	+	+	+	+	+	+	+	–	+	
E. gergoviae	–	(+)	–	+w	+	–	–	+	+	+	–	–	–	+
E. mori	+	+	ND	ND	+	ND	+	ND	+	ND	ND	ND	ND	+
E. bugandensis	–	–	+	–	+w	+	+	ND	+	+	ND	+	–	ND

Data are from references 4, 5, 12, 48. Abbreviations: LDC, lysine decarboxylase; ADH, arginine dihydrolase; URE, urease; ESC, esculin; INO, inositol; SOR, sorbitol; SAC, saccharose; MEL, melibiose; RAF, raffinose; AMG α-methyl-D-glucoside; ADO, adonitol; ARL, D-arabitol. +, positive reaction (>,90% of strains); (+), generally positive reaction; +w, weakly positive reaction; V, variable reaction; (–), generally negative reaction; –, negative reaction (<90% of strains); +/–, variable depending on the method used; ND, not determined.

5. Species Variations of Genus *Enterobacter*

5.1. *Enterobacter amnigenus* (Biogroup)

Enterobacter amnigenus, was described in 1981 by Izard et al., [42] is a rarely isolated species. As suggested by Brady et al., this bacterium was reclassified in the genus *Lelliottia* based on multilocus sequence analysis. However, that change is never been validly published, and till yet, *E. amnigenus* still remains the official nomenclature [35]. It comprises of genotypically and phenotypically different groups that have been called biogroup 1 and biogroup 2. The strains show *ornithine decarboxylase* (ODC) positive and lysine decarboxylase (LDC) and urease negative test and ferment melibiose. Strains of *E. amnigenus* (biogroup 1) ferments sucrose and raffinose but not the D-sorbitol. These are arginine dihydrolase (ADH) negative. Strains of the *E. amnigenus* (biogroup 2) ferment D-sorbitol but not sucrose and raffinose (Table 1) [43,–45].

5.2. *Enterobacter cancerogenus*

Previously known as *Erwinia cancerogena*, this species was re-named in 1988 to the genus *Enterobacter* after DNA-DNA hybridizations technique studies [46]. In 1989, Grimont and Ageron reported the synonymy between *E. cancerogenus* and *E. taylorae* (46). Now, both denominations are still valid. Nevertheless, the name *E. cancerogenus* should be retained because it was established earlier. These strains are ODC and ADH positive and LDC and urease negative (Table 1) [47]. They have an inducible chromosomal AmpC β-lactamase [45, 48].

5.3. *Enterobacter cloacae* Complex

Enterobacter cloacae complex includes *E. asburiae*, *E. carcinogenus*, *E. cloacae*, *E. hormaechei*, *E. kobei*, *E. nimipressuralis*, and *E. mori*. All these species are genotypically very close, with more than 60% DNA-DNA homology or similarity. A phylogenetic study by Hofmann and Roggenkamp, based on sequences of four housekeeping genes, confirmed the genetic diversity of the *Enterobacter cloacae* complex, species form distinct clusters [3]. *E. cloacae* and *E. hormaechei* are the most commonly found in human clinical specimens.

5.4. *Enterobacter asburiae*

E. asburiae was discovered in 1986 by Brenner et al., [5] from strains of the enteric group. It is sometimes described as *Enterobacter muelleri*. These strains are sometimes non-motile, having a Voges-Proskauer (VP)-negative test. These are indole negative, ferment D-sorbitol and sucrose, and do not ferment melibiose. When these are VP positive, it is important to differentiate them from *E. cloacae* (*E. asburiae* does not have ADH and does not ferment L-rhamnose) and other VP-positive species (*E. asburiae* does not possess LDC, Tween 80 esterase, or DNase, characteristics possessed by *Serratia marcescens* and *Serratia liquefaciens*). *E. asburiae* has occasionally been designated as a bacterium with clinical significance. It is mainly found in blood cultures. de Florio et al., studied its gradual increase in 2017 [51]. The complete genome of a clinical *E. asburiae* isolate from bone marrow transplant patient sample has been sequenced [52].

5.5. *Enterobacter cloacae*

E. cloacae is the "type species" of the genus *Enterobacter* and is divided into two subspecies i.e., *E. cloacae* subsp. *cloacae*, (esculin test is negative), and *E. cloacae* subsp. *dissolvens*, with positive esculin reaction [55]. The strains are ODC and ADH positive and LDC negative (among the bacteria of the genus *Enterobacter*, only the species *E. cloacae*, *E. taylorae*, *E. kobei*, and possibly *E. amnigenus* have this profile of decarboxylases). They ferment D-sorbitol, sucrose, and melibiose. Additional characteristics are presented in Table 1.

5.6. *Enterobacter hormaechei*

E. hormaechei was described by O'Hara et al., [55] in 1989. These bacteria are LDC- and gelatinase-negative strains which are ODC, ADH, and urease positive, and they ferment sucrose and L-rhamnose but not D-sorbitol or melibiose. These characteristics generally make it possible to differentiate this entity from phenotypically close species. Hoffmann et al., divided *E. hormaechei* into three subspecies based on biochemical results for D-adonitol, D-arabitol, D-sorbitol, D-melibiose, and dulcitol: *E. hormaechei* subsp. *oharae*, which ferments only melibiose; *E. hormaechei* subsp. *hormaechei*, (ferments only dulcitol); and *E. hormaechei* subsp. *steigerwaltii*, (ferments all of these compounds except for dulcito)l. Two more subspecies have now been characterized by whole-genome comparisons and average nucleotide identity from complete genome sequencing: *E. hormaechei* subsp. *xiangfangensis* and *E. hormaechei* subsp. *hoffmannii* [41].

5.7. *Enterobacter kobei*

Enterobacter kobei name was proposed by Kosako et al., which was previously included in *E. cloacae* [55]. These bacteria differ from *E. cloacae* by a negative VP reaction.

5.8. *Enterobacter ludwigii*

The new species *E. ludwigii* has been studied on the basis of genotypic and phenotypic properties [37]. It is genetically similar to *E. hormaechei*, and a Biotype 100 Gallery API system was used to differentitae it from other *Enterobacter* spp. by its ability to use *myo*-inositol and methyl-Dglucopyranose. Whole-genome sequencing of the type was conducted. The strains are ADH and ODC positive and LDC and urease negative.

5.9. *Enterobacter mori*

Enterobacter mori was first explained as a phyto-pathogenic bacterium. It was isolated from infected mulberry roots. The species can be differentiated from closely related other species as it shows lysine decarboxylase activity and has the ability to use D-arabitol. The type bacterium R18-2T (= CGMCC 1.10322T = LMG 25706T) was sequenced. The 16S rRNA gene sequence and MLSA indicated that *E. mori* is closely related to *E. asburiae* (*E. muelleri*) [4]. Sixty-six genes potentially involved in the secretion system have been identified, explaining the phytopathogenic nature of this species [39].

5.10. *Enterobacter nimipressuralis*

Enterobacter nimipressuralis was first studied as *Erwinia* sp. and was reclassified as *Enterobacter* in 1986 by Brenner et al., [5] on the basis of DNA-DNA hybridizations. Strains of *E. nimipressuralis* are non pigmented and are generally ADH positive and urease negative. These bacteria ferment D-sorbitol and melibiose but not sucrose. These properties most often help to differentiate these from other species that are morphologically similar to the genus *Enterobacter*. Hoffmann cluster X is *E. nimipressuralis*, and it has been reclassified as *Lelliottia nimipressuralis* by MLSA [3].

5.11. *Enterobacter aerogenes*

Enterobacter aerogenes is genotypically and morphologically similar to a motile *K. pneumoniae* strain because of its peritrichous, ODC-positive, urease-negative, and indole-negative flagella (Table 1). Izard et al., gave suggestions to reclassify *E. aerogenes* as *Klebsiella* under the name *K. mobilis*. While it is taxonomically characterized (by DNA-DNA hybridization). This proposal has not yet been accepted by medical microbiologists, who maintain the name *E. aerogenes*. From genome sequencing performed on a resistant clinical isolate, Diene et al., [37] have recently suggested the shift of the *E. aerogenes* species to the genus *Klebsiella* (*K. aeromobilis)*. Several genes involved in the bacterial motility which could have been borrowed from the *Serratia* genus, and the conjugative plasmid also could have been constructed from various transposons [37].

5.12. *Enterobacter gergoviae*

Enterobacter gergoviae was first time studied by Richard et al., in 1976 by observing a multidrug-resistant hospital strain isolated in Gergòvia near Clermont-Ferrand, France [11]. Its nomenclature was confirmed in 1980 by Brenner et al., after a DNA-DNA hybridization study [5]. Recently, it is suggested to include this species in the genus *Pluribacter* as *P. gergoviae*, based on the sequence analysis of four genes according to MLSA [35]. These bacteria are generally LDC and ODC positive and gelatinase negative. They are urease positive and do not ferment inositol, D-sorbitol, and mucate, characteristics that differentiate them from *E. aerogenes*. Unlike other *Enterobacter* bacteria, *E. gergoviae* does not grow in potassium cyanide broth.

6. Recent Molecular Techniques for Species Nomenclature/Characterization

Enterobacter bugandensis, *E. timonensis*, *E. massiliensis*, *E. chengduensis*, *E. sichuanensis*, and *E. roggenkampii* were recently described based on:

1. a computational analysis of sequenced *Enterobacter* genomes
2. MLSA of housekeeping genes [20, 40, 41].
3. *E. timonensis* and *E. massiliensis* were described on the basis of mass spectrometry (MS) and 16S rRNA DNA-DNA hybridization [41].

References

[1] Hormaeche E, Edwards P. (1960). A proposed genus *Enterobacter*. *Int Bull Bacteriol Nomencl Taxon* 10:71–74. doi:10.1099/0096266X-10-2-71. [CrossRef] [Google Scholar].

[2] Iversen C, Mullane N, McCardell B, Tall BD, Lehner A, Fanning S, Stephan R, Joosten H. (2008). *Cronobacter* gen. nov., a new genus to accommodate the biogroups of *Enterobacter sakazakii*, and proposal of *Cronobacter sakazakii* gen. nov., comb. nov., *Cronobacter malonaticus* sp. nov., *Cronobacter turicensis* sp. nov., *Cronobacter muytjensii* sp. nov., *Cronobacter dublinensis* sp. nov., *Cronobacter* genomospecies 1, and of three subspecies, *Cronobacter dublinensis* subsp. dublinensis subsp. nov., *Cronobacter dublinensis* subsp. lausannensis subsp. nov. and *Cronobacter dublinensis* subsp. lactaridi subsp. nov. *Int J Syst Evol Microbiol* 58:1442–1447. doi:10.1099/ijs.0.65577-0. [PubMed] [CrossRef] [Google Scholar].

[3] Hoffmann H, Roggenkamp A. (2003). Population genetics of the Nomenspecies *Enterobacter cloacae*. *Appl Environ Microbiol* 69:5306–5318. doi:10.1128/AEM.69.9.5306-5318.2003. [PMC free article] [PubMed] [CrossRef] [Google Scholar].

[4] Singh NK, Bezdan D, Checinska Sielaff A, Wheeler K, Mason CE, Venkateswaran K. (2018). Multi-drug resistant *Enterobacter bugandensis* species isolated from the International Space Station and comparative genomic analyses with human pathogenic strains. *BMC Microbiol* 18:175. doi:10.1186/s12866-018-1325-2. [PMC free article] [PubMed] [CrossRef] [Google Scholar].

[5] Brenner DJ, McWorther AC, Kai A, Steigerwalt AG, Farmer JJ. (1986). *Enterobacter asburiae* sp nov, a new species found in clinical specimens, and reassignment of Erwinia dissolvens and Erwinia nimipressuralis to the genus *Enterobacter* as *Enterobacter* dissolvens comb. nov. and *Enterobacter nimipressuralis* comb. nov. *J Clin Microbiol* 23:1114–1120. [PMC free article] [PubMed] [Google Scholar].

[6] Zhuang L, Zhou S, Yuan Y, Liu T, Wu Z, Cheng J. (2011). Development of *Enterobacter aerogenes* fuel cells: from in situ biohydrogen oxidization to direct electroactive biofilm. *Bioresour Technol* 102:284–289. doi:10.1016/j.biortech.2010.06.038. [PubMed] [CrossRef] [Google Scholar].

[7] Akbari M, Bakhshi B, Najar Peerayeh S. (2016). Particular distribution of *Enterobacter cloacae* strains isolated from urinary tract infection within clonal complexes. *Iran Biomed J* 20:49–55. [PMC free article] [PubMed] [Google Scholar].

[8] Bertrand X, Hocquet D, Boisson K, Siebor E, Plésiat P, Talon D. (2003). Molecular epidemiology of Enterobacteriaceae producing extended-spectrum β-lactamase in a French university-affiliated hospital. *Int J Antimicrob Agents* 22:128–133. doi:10.1016/S0924-8579(03)00098-0. [PubMed] [CrossRef] [Google Scholar].

[9] Chow JW, Fine MJ, Shlaes DM, Quinn JP, Hooper DC, Johnson MP, Ramphal R, Wagener MM, Miyashiro DK, Yu VL. (1991). *Enterobacter bacteremia*: clinical features and emergence of antibiotic resistance during therapy. *Ann Intern Med* 115:585–590. doi:10.7326/0003-4819-115-8-585. [PubMed] [CrossRef] [Google Scholar].

[10] Paauw A, Caspers MP, Leverstein-van Hall MA, Schuren FH, Montijn RC, Verhoef J, Fluit AC. (2009). Identification of resistance and virulence factors in an epidemic *Enterobacter hormaechei* outbreak strain. *Microbiology* 155:1478–1488. doi:10.1099/mic.0.024828-0. [PubMed] [CrossRef] [Google Scholar].

[11] Morand PC, Billoet A, Rottman M, Sivadon-Tardy V, Eyrolle L, Jeanne L, Tazi A, Anract P, Courpied JP, Poyart C, Dumaine V.

(2009). Specific distribution within the *Enterobacter cloacae* complex of strains isolated from infected orthopedic implants. *J Clin Microbiol* 47:2489–2495. doi:10.1128/JCM.00290-09. [PMC free article] [PubMed] [CrossRef] [Google Scholar].

[12] Sanders WE, Sanders CC. (1997). *Enterobacter* spp.: pathogens poised to flourish at the turn of the century. *Clin Microbiol Rev* 10:220–241. doi:10.1128/CMR.10.2.220. [PMC free article] [PubMed] [CrossRef] [Google Scholar].

[13] Allerberger F, Koeuth T, Lass-Florl C, Dierich MP, Putensen C, Schmutzhard E, Mohsenipour I, Grundmann H, Hartung D, Bauernfeind A, Eberlein E, Lupski JR. (1996). Epidemiology of infections due to multiresistant *Enterobacter aerogenes* in a university hospital. *Eur J Clin Microbiol Infect Dis* 15:517–521. doi:10.1007/ BF01691323. [PubMed] [CrossRef] [Google Scholar].

[14] Boucher HW, Talbot GH, Bradley JS, Edwards JE, Gilbert D, Rice LB, Scheld M, Spellberg B, Bartlett J. (2009). Bad bugs, no drugs: no ESKAPE! An update from the Infectious Diseases Society of America. *Clin Infect Dis* 48:1–12. doi:10.1086/595011. [PubMed] [CrossRef] [Google Scholar].

[15] Canton R, Oliver A, Coque TM, Varela M. d C, Perez-Diaz JC, Baquero F. (2002). Epidemiology of extended-spectrum β-lactamase producing *Enterobacter* isolates in a Spanish hospital during a 12-years period. *J Clin Microbiol* 40:1237–1243. doi:10.1128/JCM.40. 4.1237-1243.2002. [PMC free article] [PubMed] [CrossRef] [Google Scholar].

[16] Davin-Regli A, Monnet D, Saux P, Bosi C, Charrel RN, Barthelemy A, Bollet C. (1996). Molecular epidemiology of *Enterobacter aerogenes* acquisition: one-year prospective study in two intensive care units. *J Clin Microbiol* 34:1474–1480. [PMC free article] [PubMed] [Google Scholar].

[17] Davin-Regli A, Bosi C, Charrel RN, Ageron E, Papazian L, Grimont PAD, Cremieux A, Bollet C. (1997). A nosocomial outbreak due to *Enterobacter cloacae* strains with the *E. hormachei* genotype in

patients treated with fluoroquinolone treatement. *J Clin Microbiol* 35:1008–1010. [PMC free article] [PubMed] [Google Scholar].

[18] De Champs C, Sauvant MP, Chanal C, Sirot D, Gazuy N, Malhuret R, Baguet JC, Sirot J. (1989). Prospective survey of colonization and infection caused by expanded spectrum-beta-lactamase-producing members of the family Enterobacteriaceae in an intensive care unit. *J Antimicrob Chemother* 27:2887–2890. [PMC free article] [PubMed] [Google Scholar].

[19] Wenger PN, Tokars JI, Brennan P, Samel C, Bland L, Miller M, Carson L, Arduino M, Edelstein P, Aguero S, Riddle C, O'Hara C, Jarvis W. (1997). An outbreak of *Enterobacter hormachei* infection and colonization in an intensive care nursery. *Clin Infect Dis* 24:1243–1244. doi:10.1086/513650. [PubMed] [CrossRef] [Google Scholar].

[20] Wu W, Feng Y, Zong Z. (2018). *Enterobacter sichuanensis* sp. nov., recovered from human urine. *Int J Syst Evol Microbiol* 68:3922–3927. doi:10.1099/ijsem.0.003089. [PubMed] [CrossRef] [Google Scholar].

[21] Anastay M, Lagier E, Blanc V, Chardon H. (2013). Epidémiologie des bêtalactamases à spectre étendu (BLSE) chez les entérobactéries dans un hôpital du sud de la France [Extended spectrum beta-lactamase (ESBL) epidemiology in enterobacteriaceae in a hospital in the south of France], 1997–2007. *Path Biol* 61:38–43. doi:10.1016/j.patbio.2012.03.001. [PubMed] [CrossRef] [Google Scholar].

[22] Arpin C, Coze C, Rogues AM, Gachie JP, Bebear C, Quentin C. (1996). Epidemiological study of an outbreak due to multidrug-resistant *Enterobacter aerogenes* in a medical intensive care unit. *J Clin Microbiol* 34:2163–2169. [PMC free article] [PubMed] [Google Scholar].

[23] Bosi C, Davin-Regli A, Bornet C, Mallea M, Pages JM, Bollet C. (1999). Most *Enterobacter aerogenes* strains in France belong to a prevalent clone. *J Clin Microbiol* 37:2165–2169. [PMC free article] [PubMed] [Google Scholar].

[24] Davin-Regli A, Pagès JM. (2015). *Enterobacter aerogenes* and *Enterobacter cloacae*; versatile bacterial pathogens confronting antibiotic treatment. *Front Microbiol* 6:392. doi:10.3389/fmicb.2015.00392. [PMC free article] [PubMed] [CrossRef] [Google Scholar].

[25] Davin-Regli A, Masi M, Bialek S, Nicolas-Chanoine MH, Pagès JM. (2016). Antimicrobial resistance and drug efflux pumps in *Enterobacter* and *Klebsiella*, p 281–306. In Li X-Z, Elkins CA, Zgurskaya HI (ed), *Efflux-mediated drug resistance in bacteria: mechanisms, regulation and clinical implications*. Springer International Publishing, Basel, Switzerland. [Google Scholar].

[26] Chollet R, Bollet C, Chevalier J, Malléa M, Pagès JM, Davin-Regli A. (2002). mar operon involved in multidrug resistance of *Enterobacter aerogenes*. *Antimicrob Agents Chemother* 46:1093–1097. doi:10.1128/aac.46.4.1093-1097.2002. [PMC free article] [PubMed] [CrossRef] [Google Scholar].

[27] Chollet R, Chevalier J, Bryskier A, Pagès JM. (2004). The AcrAB-TolC pump is involved in macrolide resistance but not in telithromycin efflux in *Enterobacter aerogenes* and *Escherichia coli*. *Antimicrob Agents Chemother* 48:3621–3624. doi:10.1128/AAC.48.9.3621-3624.2004. [PMC free article] [PubMed] [CrossRef] [Google Scholar].

[28] Dam S, Pagès JM, Masi M. (2018). Stress responses, outer membrane permeability control and antimicrobial resistance in Enterobacteriaceae. *Microbiology* 164:260–267. doi:10.1099/mic.0.000613. [PubMed] [CrossRef] [Google Scholar].

[29] Masi M, Réfregiers M, Pos KM, Pagès JM. (2017). Mechanisms of envelope permeability and antibiotic influx and efflux in Gram-negative bacteria. *Nat Microbiol* 2:17001. doi:10.1038/nmicrobiol.2017.1. [PubMed] [CrossRef] [Google Scholar].

[30] Molitor A, James CE, Fanning S, Pagès JM, Davin-Regli A. (2018). Ram locus is a key regulator to trigger multidrug resistance in *Enterobacter aerogenes*. *J Med Microbiol* 67:148–159. doi:10.1099/jmm.0.000667. [PubMed] [CrossRef] [Google Scholar].

[31] Pérez A, Poza M, Fernández A, del Carmen Fernández M, Mallo S, Merino M, Rumbo-Feal S, Cabral MP, Bou G. (2012). Involvement of the AcrAB-TolC efflux pump in the resistance, fitness, and virulence of *Enterobacter cloacae*. *Antimicrob Agents Chemother* 56:2084–2090. doi:10.1128/AAC.05509-11. [PMC free article] [PubMed] [CrossRef] [Google Scholar].

[32] Izdebski R, Baraniak A, Herda M, Fiett J, Bonten MJM, Carmeli Y, Goossens H, Hryniewicz W, Brun-Buisson C, Gniadkowski M, Grabowska A, Nikonorow E, Derde LPG, Dautzenberg MJ, Adler A, Kazma M, Navon-Venezia S, Malhotra-Kumar S, Lammens C, Dumpis U, Giamarellou H, Muzlovic I, Nardi G, Petrikkos GL, Stammet P, Salomon J, Lawrence C, Legrand P, Rossini A, Salvia A, Samso JV, Fierro J, Paul M, Lerman Y, MOSAR WP2, WP3 and WP5 Study Groups. (2015). MLST reveals potentially high-risk international clones of *Enterobacter cloacae*. *J Antimicrob Chemother* 70:48–56. doi:10.1093/jac/dku359. [PubMed] [CrossRef] [Google Scholar].

[33] Liu WY, Wong CF, Chung KM, Jiang JW, Leung FC. (2013). Comparative genome analysis of *Enterobacter cloacae*. *PLoS One* 8:e74487. doi:10.1371/journal.pone.0074487. [PMC free article] [PubMed] [CrossRef] [Google Scholar].

[34] Paauw A, Caspers MPM, Schuren FHJ, Hall MAL, Delétoile A, Montijn RC, Verhoef J, Fluit AC. (2008). Genomic diversity within the *Enterobacter cloacae* complex. *PLoS One* 3:e3018. doi:10.1371/journal.pone.0003018. [PMC free article] [PubMed] [CrossRef] [Google Scholar].

[35] Brady C, Cleenwerck I, Venter S, Coutinho T, De Vos P. (2013). Taxonomic evaluation of the genus *Enterobacter* based on multilocus sequence analysis (MLSA): proposal to reclassify *E. nimipressuralis* and *E. amnigenus* into Lelliottia gen. nov. as *Lelliottia nimipressuralis* comb. nov. and *Lelliottia amnigena* comb. nov., respectively, *E. gergoviae* and *E. pyrinus* into Pluralibacter gen. nov. as *Pluralibacter gergoviae* comb. nov. and *Pluralibacter pyrinus* comb. nov., respectively, *E. cowanii*, *E. radicincitans*, *E. oryzae* and

E. arachidis into Kosakonia gen. nov. as *Kosakonia cowanii* comb. nov., *Kosakonia radicincitans* comb. nov., *Kosakonia oryzae* comb. nov. and *Kosakonia arachidis* comb. nov., respectively, and *E. turicensis*, *E. helveticus* and *E. pulveris* into Cronobacter as *Cronobacter zurichensis* nom. nov., *Cronobacter helveticus* comb. nov. and *Cronobacter pulveris* comb. nov., respectively, and emended description of the genera *Enterobacter and Cronobacter*. *Syst Appl Microbiol* 36:309–319. doi:10.1016/j.syapm.2013.03.005. [PubMed] [CrossRef] [Google Scholar].

[36] Izard D, Gavini F, Leclerc H. (1980). Polynucleotide sequence relatedness and genome size among *Enterobacter* intermedium sp nov and the species *Enterobacter cloacae* and *Klebsiella pneumoniae*. *Zentralbl Bakteriol Parasitenkd Infektionskr Hyg Abt 1 Orig Reihe C* 1:51–60. doi:10.1016/S0172-5564(80)80016-9. [CrossRef] [Google Scholar].

[37] Diene SM, Merhej V, Henry M, El Filali A, Roux V, Robert C, Azza S, Gavory F, Barbe V, La Scola B, Raoult D, Rolain JM. (2013). The rhizome of the multidrug-resistant *Enterobacter aerogenes* genome reveals how new "killer bugs" are created because of a sympatric lifestyle. *Mol Biol Evol* 30:369–383. doi:10.1093/molbev/mss236. [PubMed] [CrossRef] [Google Scholar].

[38] Tindall BJ, Sutton G, Garrity GM. 2017. *Enterobacter aerogenes* Hormaeche and Edwards 1960 (Approved Lists 1980) and *Klebsiella mobilis* Bascomb et al., 1971 (Approved Lists 1980) share the same nomenclatural type (ATCC 13048) on the Approved Lists and are homotypic synonyms, with consequences for the name *Klebsiella mobilis* Bascomb et al., 1971 (Approved Lists 1980). *Int J Syst Evol Microbiol* 67:502–504. doi:10.1099/ijsem.0.001572. [PubMed] [CrossRef] [Google Scholar].

[39] Hartl R, Kerschner H, Gattringer R, Lepuschitz S, Allerberger F, Sorschag S, Ruppitsch W, Apfalter P. (2018). Whole-genome analysis of a human *Enterobacter mori* isolate carrying a bla_{IMI-2} carbapenemase in Austria. *Microb Drug Resist* doi:10.1089/mdr.2018.0098. [PubMed] [CrossRef] [Google Scholar].

[40] Tidjani Alou M, Cadoret F, Brah S, Diallo A, Sokhna C, Mehrej V, Lagier JC, Fournier PE, Raoult D. (2017). '*Khelaifiella massiliensis*,' '*Niameybacter massiliensis*,' '*Brachybacterium massiliense*,' '*Enterobacter timonensis*,' '*Massilibacillus massiliensis*,' new bacterial species and genera isolated from the gut microbiota of healthy infants. *New Microbes New Infect* 19:1–7. doi:10.1016/ j.nmni.2017.02.002. [PMC free article] [PubMed] [CrossRef] [Google Scholar].

[41] Sutton GG, Brinkac LM, Clarke TH, Fouts DE. (2018). *Enterobacter hormaechei* subsp. *hoffmannii* subsp. nov., *Enterobacter hormaechei* subsp. xiangfangensis comb. nov., *Enterobacter roggenkampii* sp. nov., and *Enterobacter muelleri* is a later heterotypic synonym of *Enterobacter asburiae* based on computational analysis of sequenced *Enterobacter genomes. F1000 Res* 7:521. doi:10.12688/f1000 research.14566.1. [PMC free article] [PubMed] [CrossRef] [Google Scholar].

[42] Izard D, Gavini F, Trinel PA, Leclerc H. (1981). Desoxyribonucleic acid relatedness between *Enterobacter cloacae* and *Enterobacter amnigenus* sp nov. *Int J Syst Bacteriol* 31:35–42. doi:10.1099/ 00207713-31-1-35. [CrossRef] [Google Scholar].

[43] Bollet C, Elkouby A, Pietri P, Micco P. (1991). *Enterobacter amnigenus* isolated from a heart transplant recipient. *Eur J Clin Microbiol Infect Dis* 10:1071–1073. doi:10.1007/BF01984933. [PubMed] [CrossRef] [Google Scholar].

[44] Freney J, Husson MO, Gavini F, Madier S, Martra A, Izard D, Leclerc H, Fleurette J. (1988). Susceptibility to antibiotics and antiseptics of new species of the family Enterobacteriaceae. *Antimicrob Agents Chemother* 32:873–876. doi:10.1128/aac. 32.6.873. [PMC free article] [PubMed] [CrossRef] [Google Scholar].

[45] Muytjens HL, van der Ros-van de Repe J. (1986). Comparative in vitro susceptibilities of eight *Enterobacter* species, with special reference to *Enterobacter sakazakii. Antimicrob Agents Chemother* 29:367–370. doi:10.1128/aac.29.2.367. [PMC free article] [PubMed] [CrossRef] [Google Scholar].

[46] Grimont PAD, Ageron E. (1989). *Enterobacter cancerogenus* (Urosevic, 1966) Dickey and Zumoff 1988, a senior subjective synonym of *Enterobacter taylorae* Farmer et al., (1985). *Res Microbiol* 140:459–465. doi:10.1016/0923-2508(89)90067-3. [PubMed] [CrossRef] [Google Scholar].

[47] Stock I, Wiedemann B. (2002). Natural antibiotic susceptibility of *Enterobacter amnigenus*, *Enterobacter cancerogenus*, *Enterobacter gergoviae* and *Enterobacter sakasakii*. *Clin Microbiol Infect* 8:564–578. doi:10.1046/j.1469-0691.2002.00413.x. [PubMed] [CrossRef] [Google Scholar].

[48] Farmer JJ III, Fanning GR, Davis BR, O'Hara CM, Riddle C, Hickman-Brenner FW, Asbury MA, Lowery VA III, Brenner DJ. (1985). *Escherichia fergusonii* and *Enterobacter taylorae*, two new species of Enterobactriaceae isolated from clinical specimens. *J Clin Microbiol* 21:77–81. [PMC free article] [PubMed] [Google Scholar].

[49] Freney J, Gavini F, Ploton C, Leclerc H, Fleurette J. (1986). Isolement d'une souche d'*Enterobacter taylorae* chez un brûlé [Isolation of a strain of *Enterobacter taylorae* from a burn victim]. *Méd Mal Infect* 5:320–321. doi:10.1016/S0399-077X(86)80248-7. [CrossRef] [Google Scholar].

[50] Westblom TU, Coggins ME. (1987). Osteomyelitis caused by *Enterobacter taylorae*, formerly enteric group 19. *J Clin Microbiol* 25:2432–2433. [PMC free article] [PubMed] [Google Scholar].

[51] De Florio L, Riva E, Giona A, Dedej E, Fogolari M, Cella E, Spoto S, Lai A, Zehender G, Ciccozzi M, Angeletti S. (2018). MALDI-TOF MS identification and clustering applied to *Enterobacter* species in nosocomial setting. *Front Microbiol* 9:1885. doi:10.3389/fmicb. 2018.01885. [PMC free article] [PubMed] [CrossRef] [Google Scholar].

[52] Zhu B, Wang S, Li O, Hussain A, Hussain A, Shen J, Ibrahim M. (2017). High-quality genome sequence of human pathogen *Enterobacter asburiae* type strain 1497-78^{T}. *J Glob Antimicrob Resist* 8:104–105. doi:10.1016/j.jgar.2016.12.003. [PubMed] [CrossRef] [Google Scholar].

[53] Aubron C, Poirel L, Ash RJ, Nordmann P. (2005). Carbapenemase-producing Enterobacteriaceae, US rivers. *Emerg Infect Dis* 11:260–264. doi:10.3201/eid1102.030684. [PMC free article] [PubMed] [CrossRef] [Google Scholar].

[54] Rotova V, Papagiannitsis CC, Chudejova K, Medvecky M, Skalova A, Adamkova V, Hrabak J. (2017). First description of the emergence of *Enterobacter asburiae* producing IMI-2 carbapenemase in the Czech Republic. *J Glob Antimicrob Resist* 11:98–99. doi:10.1016/j.jgar.2017.10.001. [PubMed] [CrossRef] [Google Scholar].

[55] Hoffmann H, Stindl S, Ludwig W, Stumpf A, Mehlen A, Heesemann J, Monget D, Schleifer KH, Roggenkamp A. (2005). Reassignment of *Enterobacter* dissolvens to *Enterobacter cloacae* as *E. cloacae* subspecies dissolvens comb. nov. and emended description of *Enterobacter asburiae* and *Enterobacter kobei*. *Syst Appl Microbiol* 28:196–205. doi:10.1016/j.syapm.2004.12.010. [PubMed] [CrossRef] [Google Scholar].

INDEX

A

B

C

D

E

F

G

H

K

L

M

N

O

P

R

S

T

U

V

W

Y